Lecture Notes in Computer Science   11855

More information about this series at http://www.springer.com/series/7412

Huazhu Fu · Mona K. Garvin ·
Tom MacGillivray · Yanwu Xu ·
Yalin Zheng (Eds.)

Ophthalmic Medical Image Analysis

6th International Workshop, OMIA 2019
Held in Conjunction with MICCAI 2019
Shenzhen, China, October 17
Proceedings

Springer

Editors
Huazhu Fu (iD)
Inception Institute
of Artificial Intelligence
Abu Dhabi, United Arab Emirates

Tom MacGillivray (iD)
University of Edinburgh
Edinburgh, UK

Yalin Zheng (iD)
The University of Liverpool
Liverpool, UK

Mona K. Garvin (iD)
University of Iowa
Iowa City, IA, USA

Yanwu Xu (iD)
Baidu, Inc.
Beijing, China

ISSN 0302-9743   ISSN 1611-3349 (electronic)
Lecture Notes in Computer Science
ISBN 978-3-030-32955-6   ISBN 978-3-030-32956-3 (eBook)
https://doi.org/10.1007/978-3-030-32956-3

LNCS Sublibrary: SL6 – Image Processing, Computer Vision, Pattern Recognition, and Graphics

This Springer imprint is published by the registered company Springer Nature Switzerland AG
The registered company address is: Gewerbestrasse 11, 6330 Cham, Switzerland

Preface

The 6th International Workshop on Ophthalmic Medical Image Analysis (OMIA 2019) was held in Shenzhen, China, on October 17, 2019, in conjunction with the 22nd International Conference on Medical Image Computing and Computer-Assisted Intervention (MICCAI).

Age-related macular degeneration, diabetic retinopathy, and glaucoma are the main causes of blindness in both developed and developing countries. The cost of blindness to society and individuals is huge, and many cases can be avoided by early intervention. Early and reliable diagnosis strategies and effective treatments are therefore a world priority. At the same time, there is mounting research on the retinal vasculature and neuro-retinal architecture as a source of biomarkers for several high-prevalence conditions like dementia, cardiovascular disease, and of course complications of diabetes. Automatic and semi-automatic software tools for retinal image analysis are being used widely in retinal biomarkers research, and increasingly percolating into clinical practice. Significant challenges remain in terms of reliability and validation, number and type of conditions considered, multi-modal analysis (e.g., fundus, optical coherence tomography, scanning laser ophthalmoscopy), novel imaging technologies, and the effective transfer of advanced computer vision and machine learning technologies, to mention a few. The workshop addressed all these aspects and more, in the ideal interdisciplinary context of MICCAI.

This workshop aimed to bring together scientists, clinicians, and students from multiple disciplines in the growing ophthalmic image analysis community, such as electronic engineering, computer science, mathematics, and medicine, to discuss the latest advancements in the field. A total of 36 full-length papers were submitted to the workshop in response to the call for papers. All submissions are double-blind peer-reviewed by at least three members of the Program Committee. Paper selection was based on methodological innovation, technical merit, results, validation, and application potential. Finally, 22 papers were accepted for orals (8 papers) and posters (14 papers) at the workshop and chosen to be included in this Springer LNCS volume.

We are grateful to the Program Committee for reviewing the submitted papers and giving constructive comments and critiques, to the authors for submitting high-quality papers, to the presenters for excellent presentations, and to all the OMIA 2019 attendees for coming to Shenzhen from all around the world.

September 2019

Huazhu Fu
Mona K. Garvin
Tom MacGillivray
Yanwu Xu
Yalin Zheng

Organization

Workshop Organizers

Huazhu Fu   Inception Institute of Artificial Intelligence, UAE
Mona K. Garvin   University of Iowa, USA
Tom MacGillivray   University of Edinburgh, UK
Yanwu Xu   Baidu Inc., China
Yalin Zheng   University of Liverpool, UK

Program Committee

Bhavna Antony   IBM Research, Australia
Guozhen Chen   Shenzhen University, China
Min Chen   University of Pennsylvania, USA
Qiang Chen   Nanjing University of Science and Technology, China
Jun Cheng   Cixi Institute of Biomedical Engineering, Chinese
                          Academy of Sciences, China
Dongxu Gao   University of Liverpool, UK
Mohammad Hamghalam   Shenzhen University, China
Baiying Lei   Shenzhen university, China
Xiaomeng Li   The Chinese University of Hong Kong, SAR China
Dwarikanath Mahapatra   IBM Research, Australia
Emma Pead   University of Edinburgh, UK
Suman Sedai   IBM Research, Australia
Fei Shi   Soochow University, China
Raphael Sznitman   University of Bern, Switzerland
Mingkui Tan   South China University of Technology, China
Ruwan Tennakoon   RMIT University, Australia
Jui-Kai Wang   University of Iowa, USA
Dehui Xiang   Soochow University, China
Mengdi Xu   Yitu Technology, Singapore
Yuguang Yan   South China University of Technology, China
Jiong Zhang   University of Southern California, USA
Yitian Zhao   Cixi Institute of Biomedical Engineering, Chinese
                          Academy of Sciences, China
Yuanjie Zheng   Shandong Normal University, China
Weifang Zhu   Soochow University, China

Contents

Dictionary Learning Informed Deep Neural Network with Application to OCT Images

Joshua Bridge[1], Simon P. Harding[1,2], Yitian Zhao[3], and Yalin Zheng[1,2(✉)]

[1] Department of Eye and Vision Science, University of Liverpool,
Liverpool L7 8TX, UK
{joshua.bridge,s.p.harding,yalin.zheng}@liverpool.ac.uk
[2] St. Paul's Eye Unit, Royal Liverpool University Hospital, Liverpool L7 8XP, UK
[3] Cixi Institute of Biomedical Engineering, Ningbo Institute of Industrial
Technology, Chinese Academy of Sciences, Ningbo 315201, Zhejiang, China
yitian.zhao@nimte.ac.cn

Abstract. Medical images are often of very high resolutions, far greater than can be directly processed in deep learning networks. These images are usually downsampled to much lower resolutions, likely losing useful clinical information in the process. Although methods have been developed to make the image appear much the same to human observers, a lot of information that is valuable to deep learning algorithms is lost. Here, we present a novel dictionary learning method of reducing the image size, utilizing DAISY descriptors and Improved Fisher kernels to derive features to represent the image in a much smaller size, similar to traditional downsampling methods. Our proposed method works as a type of intelligent downsampling, reducing the size while keeping vital information in images. We demonstrate the proposed method in a classification problem on a publicly available dataset consisting of 108,309 training and 1,000 validation grayscale optical coherence tomography images. We used an Inception V3 network to classify the resulting representations and to compare with previously obtained results. The proposed method achieved a testing accuracy and area under the receiver operating curve of 97.2% and 0.984, respectively. Results show that the proposed method does provide an accurate representation of the image and can be used as a viable alternative to conventional downsampling.

Keywords: Dictionary learning · Deep neural network · DAISY descriptors · Improved Fisher Kernels · OCT

1 Introduction

Routinely collected medical images are usually high resolution, far exceeding computational capabilities, meaning that these images must be downsampled to

EPSRC (2110275) and Institute of Ageing and Chronic Disease, University of Liverpool.

much lower resolutions for most deep learning applications. Any downsampling will inevitably lose information [5]. There is a need for data efficient, intelligent downsampling techniques which can downsample images while keeping as much relevant information as possible. Traditional image downsampling techniques focus on making the image appear the same to a human observer (for example, bicubic interpolation [7]). However, there is often a difference between features that humans and algorithms perceive to be important. We aim to develop a new method that is data efficient and preserves the features that a deep neural network finds useful in challenging tasks.

This paper proposes a novel method largely inspired by Albarrak et al. [1], which aims to provide a better method of image downsampling. Albarrak et al. [1] proposed a method of volumetric image classification, which involved decomposing 3D volumetric images into homogeneous regions and then representing each region with a Histogram of Oriented Gradients (HOG). They then used Improved Fisher Kernels (IFK) [13] to create one feature vector for each image. Contrary to Albarrak et al. [1], our method utilizes DAISY descriptors to provide a 2D representation of images in a 3D space. IFK is then used to choose the best value in the third dimension of the DAISY feature maps, finally resulting in a 2D image. This method results in a 2D representation, which is of a significantly smaller size than the original image. The representation can then be passed through a deep learning network for classification. The method is illustrated in Fig. 1. Our method is demonstrated on a publicly available dataset, consisting of Optical Coherence Tomography (OCT) images [6]. We compare our results with their published results obtained previously on this dataset and achieve improved accuracy. Our main contribution can be summarised as follows. First, we propose a new dictionary learning approach for intelligent image downsampling based on DAISY descriptors and Fisher Vectors; second, we demonstrate that the proposed approach is compatible with deep learning algorithms; third, we show promising results in OCT images to improve disease classification.

The remainder of the paper is organized as follows. Section 2 gives a brief outline of previous work. Section 3 describes the methods used to create the image representation. In Sect. 4, we apply our method to a dataset of OCT images and give results compared to previously obtained results. Finally, Sect. 5 briefly discusses our findings and concludes.

2   Previous Work

There are a variety of well-established image downsampling techniques, including nearest neighbor, bilinear, and bicubic interpolation [7]. Downsampling techniques mainly focus on reducing image dimensions while still providing an accurate representation of the image. Other downsampling methods, such as adaptive downsampling [9], focus on enabling the image to be reconstructed to the original size. Previously, these methods have been evaluated based upon how similar they are to the ground truth, using performance measures such as peak signal to noise ratio and root mean square error. However, these performance measures

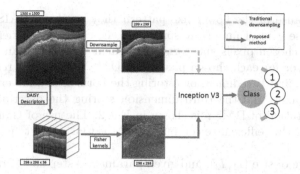

Fig. 1. The proposed framework. Previously, large images would need to be downsampled to avoid an out of memory error. The new framework aims to avoid random downsampling by introducing a more data efficient approach. The new image representation may not be more intuitive to a human observer; however, it may be more useful for a classification algorithm.

fail to evaluate how well the method captures features that may be useful in a classification task.

Methods such as Histogram of Oriented Gradients (HOG) [19], Scale Invariant Feature Transformation (SIFT) [10], dictionary learning [1], and deep learning [3], have been proposed as a method of reducing image dimension before classification. These methods have some success in producing excellent classification performance. A modified version of DAISY has previously been used in a logo classification problem [8], where it was used to produce edge maps. A linear support vector machine (SVM) was then used for classification. This method produced state-of-art results, which were superior to other SIFT-like methods. We differ from this method by using Fisher vectors to reduce the DAISY representation back to an image; this allows us to use already well-developed image classification algorithms.

3 Methods

The proposed new dictionary learning image representation consists of two steps. The first step utilizes DAISY descriptors to describe the image densely in three dimensions, creating a dictionary. The second step uses Fisher kernels to choose the most significant value in the third dimension of the DAISY features. This produces a much smaller image that resembles the original to some degree. While the generated image may not be insightful to human graders, its enhanced features may be more useful to a computer than the original image.

3.1 DAISY

DAISY is an efficient image descriptor [16], which works in a similar way to the more widely known algorithms, SIFT, and GLOH [15]. DAISY descriptors are named because of the flower like pattern they produce. DAISY begins by selecting a dense grid of uniformly spaced pixels, calculating orientation maps for each of the chosen pixels and convolving them with Gaussian kernels. This produces a vector for each chosen pixel. The DAISY algorithm results in a 3D tensor, with the first two dimensions storing the horizontal and vertical location of the chosen pixels and the third dimension storing the descriptor values. A visual representation of DAISY is shown in Fig. 2. The use of Gaussian kernels makes this algorithm efficient to compute, and it is well suited to providing dense representations [15].

For an image of size (I_1, I_2), and given parameters step size s, radius r, rings p, histograms h, and orientations o, the DAISY algorithm returns an array of size (D_1, D_2, D_3), where

$$D_1 = \left\lceil \frac{I_1 - r^2}{s} \right\rceil, \qquad D_2 = \left\lceil \frac{I_2 - r^2}{s} \right\rceil, \qquad D_3 = (p \times h + 1) \times o.$$

In our experiments, the DAISY algorithm was implemented using scikit-image 0.15.0 [17], with a step size of 5, a radius of 5, one ring, 8 histograms, and 4 orientations. These parameters were chosen after some initial testing on a small subset of 5000 images, more intensive testing may provide better parameter choices. Each direction consisted of 298 pixels, resulting in a 298 × 298 × 36 feature map.

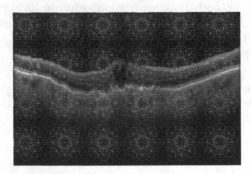

Fig. 2. Visual representation of DAISY descriptors on an example image. A sparse representation is used so that the pattern is easily observed.

3.2   Fisher Vectors

The Fisher kernel combines generative statistical models with discriminative methods, making it both generalizable and flexible [13]. It has previously been applied to a variety of problems, including image representation. Since its creation, the Fisher kernel has been improved to provide higher accuracy in real-life classification, using L2 normalization and changing the linear kernel with a non-linear additive kernel [13]. This Improved Fisher Kernel (IFK) is used here to reduce the image dimension further. In brief, for a set of feature vectors

$F = (x_1, \ldots, x_N)$ in D dimensions, such as those extracted by DAISY, we fit a Gaussian Mixture Model (GMM) with K kernels, $\Gamma = (\mu_k, \Sigma_k, \pi_k; k = 1, \ldots, k)$. The GMM calculates posterior probabilities for each feature vector:

$$p_{nk} = \frac{\exp\left[-\frac{1}{2}(x_n - \mu_k)^T \Sigma_g^{-1}(x_n - \mu_k)\right]}{\sum_{i=1}^K \exp\left[-\frac{1}{2}(x_n - \mu_i)^T \Sigma_k^{-1}(x_n - \mu_i)\right]}.$$

Then for each kernel k and dimension d, the mean and covariance are calculated:

$$m_{dk} = \frac{1}{N\sqrt{\pi_k}} \sum_{n=1}^N p_{nk} \frac{x_{dn} - \mu_{dk}}{\sigma_{dk}},$$

$$c_{dk} = \frac{1}{N\sqrt{2\pi_k}} \sum_{n=1}^N p_{dk} \left[\left(\frac{x_{dn} - \mu_{dk}}{\sigma_{dk}}\right)^2 - 1\right].$$

In order to reduce computation, we apply Fisher vectors with one kernel, giving $q_i = 1$. A Fisher Vector is calculated per image, meaning that $m_{d1} = 0, \forall d$. Hence, for our application, Fisher Vectors can be described with one vector:

$$c_{d1} = \frac{1}{N\sqrt{2\pi_1}} \sum_{n=1}^N \left[\left(\frac{x_{dn} - \mu_{d1}}{\sigma_{d1}}\right)^2 - 1\right].$$

This vector is then used to reconstruct a grayscale image. This is a special case of IFK; otherwise, we can use more kernels and which would produce a longer feature vector.

4   Experimental and Results

4.1   Data

The proposed method is demonstrated on a publicly available dataset consisting of OCT images [6]. OCT is a similar concept to ultrasound; however, it uses light instead of sound to produce a cross-sectional view of tissue composition with micrometre resolution [2]. In this dataset, each image is labeled as either normal, Choroidal Neovascularisation (CNV), Diabetic Macular Edema (DME), or drusen, corresponding to the disease that they display. CNV is a leading retinal disease that can cause irreversible sight loss. There are various causes of CNV, with the main form of CNV being wet age-related macular degeneration(AMD) [4,18]. DME, also called Diabetic Macula Oedema (DMO), is a common cause of vision loss in patients with diabetes. DME is characterized by intraretinal fluid, causing a thickening in the retinal layers [11]. Drusen are subretinal lipid deposits and are indicative of AMD [12]. Examples of each of the four classes are shown in Fig. 3.

The training dataset consists of 37,205 CNV images, 11,348 DME images, 8,616 drusen images, and 51,140 healthy images. The testing dataset comprised of 250 images from each class and was collected from patients separate from the training dataset [6]. Same as used by Kermany et al. [6] we use this as a validation dataset to evaluate performance.

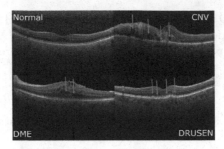

Fig. 3. Examples of the four classes in the OCT dataset. Arrows indicate the prominent features that lead to the diagnosis.

4.2 Experiments Setup

All experiments were conducted on a Linux machine using Ubuntu 18.04, with a Titan X 12 GB GPU and 32 GB of memory. Python 3.6 and Keras 2.2.4 were used to implement the method and in the deep learning network. For comparison, we followed the work of Kermany et al. [6] who used an Inception V3 network [14], pretrained on Imagenet, with the Adam optimizer to classify the images according to the disease they displayed. Inception V3 is a popular deep learning network consisting of 159 layers. The Inception V3 is based on previous Inception networks, introducing new design principles to increase accuracy while reducing computational complexity, such as. During the training of the classification model, early stopping, with a patience of 10 epochs, and model checkpoints were used to prevent overfitting and to select the best model. Class weights were used to balance the training dataset.

4.3 Results

The output of the DAISY algorithm was a $298 \times 298 \times 36$ tensor. A Fisher vector generalized mixed model was then applied in the third dimension, with kernel dimension 1. The final output resulted in a final size of 298×298, which is close to our target of 299×299. Each image took an average of 2.1 s to process offline in this way. We used the classification method described by Kermany et al. [6], which allowed us to compare our results directly with theirs.

To assess the performance of the proposed method, accuracy, macro-average multi-class area under the receiver operating characteristic (AUC), sensitivity, and specificity were calculated and compared with previously reported results by Kermany et al. [6], shown in Table 1. Our method achieved an AUC of 0.984 and outperformed [6] in terms of specificity (99.1%) and accuracy (97.2%).

5   Discussion and Conclusions

This novel method aims to reduce image size while keeping important information useful to the classification. DAISY descriptors provide a dense representation of the image and Fisher kernels further reduce the size of that representation

Table 1. Multiclass classification performance metrics in the testing dataset. Values in bold indicate the best score for that metric.

Method	Accuracy	AUC	Sensitivity	Specificity
Kermany et al. [6]	96.6%	-	**97.8%**	97.4%
Our model	**97.2%**	**0.984**	97.1%	**99.1%**

of that image. The results presented here achieve improved accuracy and specificity over the previous results by Kermany et al. [6], which utilized traditional downsampling methods. This demonstrates that the proposed method successfully captures useful information in images and may provide a better alternative to traditional downsampling methods. Our method achieves improved specificity at the expense of some sensitivity. Principal Component Analysis was used as an alternative to the Fisher kernels, however we were not able to obtain good results and these are not presented here. DAISY has the advantage of providing dense representations, while also being computationally inexpensive. The use of DAISY descriptors may also provide a more robust representation against both photometric and geometric transformations compared to other descriptor algorithms, as observed by Tola et al. [15].

The biggest current limitation of this method is the time taken to process the images; speed increases may be possible if DAISY descriptors can be calculated on a GPU. The effect of DAISY hyperparameters such as step size and radius are yet to be fully explored, and improved results may be possible. More work needs to be carried out to confirm if the proposed method generalizable to other imaging modalities such as color fundus, which contains more features than OCT.

In conclusion, we have successfully demonstrated that our new method may provide a viable alternative to downsampling images before training a deep learning network. Our method has increased accuracy over previous work when tested on a public dataset. Future work will concentrate on further optimizing the model in terms of speed and optimal parameters and in seeking other applications.

References

1. Albarrak, A., Coenen, F., Zheng, Y.: Volumetric image classification using homogeneous decomposition and dictionary learning: a study using retinal optical coherence tomography for detecting age-related macular degeneration. Comput. Med. Imaging Graph. **55**, 113–123 (2017)
2. Bezerra, H.G., Costa, M.A., Guagliumi, G., Rollins, A.M., Simon, D.I.: Intracoronary optical coherence tomography: a comprehensive review clinical and research applications. JACC. Cardiovasc. Interv. **2**(11), 1035–1046 (2009)
3. Bychkov, D., et al.: Deep learning based tissue analysis predicts outcome in colorectal cancer. Sci. Rep. **8**(1), 3395 (2018)
4. Cohen, S.Y., Laroche, A., Leguen, Y., Soubrane, G., Coscas, G.J.: Etiology of choroidal neovascularization in young patients. Ophthalmology **103**(8), 1241–1244 (1996)

5. Dodge, S.F., Karam, L.J.: Understanding how image quality affects deep neural networks. CoRR abs/1604.04004 (2016). http://arxiv.org/abs/1604.04004
6. Kermany, D.S., et al.: Identifying medical diagnoses and treatable diseases by image based deep learning. Cell **172**(5), 1122–1131.e9 (2018)
7. Keys, R.: Cubic convolution interpolation for digital image processing. IEEE Trans. Acoust. Speech Signal Process. **29**(6), 1153–1160 (1981)
8. Lei, B., Thing, V.L.L., Chen, Y., Lim, W.: Logo classification with edge-based daisy descriptor. In: 2012 IEEE International Symposium on Multimedia, pp. 222–228 (2012)
9. Lin, W., Dong, L.: Adaptive downsampling to improve image compression at low bit rates. IEEE Trans. Image Process. **15**(9), 2513–2521 (2006). https://doi.org/10.1109/TIP.2006.877415
10. Lowe, D.G.: Object recognition from local scale-invariant features. In: Proceedings of the Seventh IEEE International Conference on Computer Vision, vol. 2, pp. 1150–1157 (1999). https://doi.org/10.1109/ICCV.1999.790410
11. Otani, T., Kishi, S., Maruyama, Y.: Patterns of diabetic macular edema with optical coherence tomography. Am. J. Ophthalmol. **127**(6), 688–693 (1999)
12. Pedersen, H.R., Gilson, S.J., Dubra, A., Munch, I.C., Larsen, M., Baraas, R.C.: Multimodal imaging of small hard retinal drusen in young healthy adults. Br. J. Ophthalmol. **102**(1), 146–152 (2018)
13. Perronnin, F., Sánchez, J., Mensink, T.: Improving the fisher kernel for large-scale image classification. In: Daniilidis, K., Maragos, P., Paragios, N. (eds.) ECCV 2010. LNCS, vol. 6314, pp. 143–156. Springer, Heidelberg (2010). https://doi.org/10.1007/978-3-642-15561-1_11
14. Szegedy, C., Vanhoucke, V., Ioffe, S., Shlens, J., Wojna, Z.: Rethinking the inception architecture for computer vision. CoRR abs/1512.00567 (2015). http://arxiv.org/abs/1512.00567
15. Tola, E., Lepetit, V., Fua, P.: Daisy: an efficient dense descriptor applied to wide-baseline stereo. IEEE Trans. Pattern Anal. Mach. Intell. **32**(5), 815–830 (2010)
16. Tola, E., Lepetit, V., Fua, P.: A fast local descriptor for dense matching (2008)
17. van der Walt, S., et al.: The scikit-image contributors: scikit-image: image processing in Python. PeerJ **2**, e453 (2014). https://doi.org/10.7717/peerj.453
18. Wong, T.Y., et al.: Myopic choroidal neovascularisation: current concepts and update on clinical management. Br. J. Ophthalmol. **99**, 289–296 (2015)
19. Zhao, Y., Zhang, Y., Cheng, R., Wei, D., Li, G.: An enhanced histogram of oriented gradients for pedestrian detection. IEEE Intell. Transp. Syst. Mag. **7**(3), 29–38 (2015)

Structure-Aware Noise Reduction Generative Adversarial Network for Optical Coherence Tomography Image

Yan Guo[1], Kang Wang[2], Suhui Yang[1], Yue Wang[1], Peng Gao[1], Guotong Xie[1], Chuanfeng Lv[1], and Bin Lv[1(✉)]

[1] Ping An Technology (Shenzhen) Co., Ltd., Shenzhen, China
lvbin006@pingan.com.cn
[2] Beijing Friendship Hospital, Beijing, China

Abstract. Optical coherence tomography (OCT) is a common imaging examination in ophthalmology, which can visualize cross-sectional retinal structures for diagnosis. However, image quality still suffers from speckle noise and other motion artifacts. An effective OCT denoising method is needed to ensure the image is interpreted correctly. However, lack of paired clean image restricts its development. Here, we propose an end-to-end structure-aware noise reduction generative adversarial network (SNR-GAN), trained with un-paired OCT images. The network is designed to translate images between noisy domain and clean domain. Besides adversarial and cycle consistence loss, structure-aware loss based on structural similarity index (SSIM) is added to the objective function, so as to achieve more structural constraints during image denoising. We evaluated our method on normal and pathological OCT datasets. Compared to the traditional methods, our proposed method achieved the best denoising performance and subtle structural preservation.

Keywords: Optical coherence tomography · Image denoising · Generative adversarial network

1 Introduction

Optical coherence tomography (OCT), with its noninvasive visualization of cross-sectional tissue structures, has been regarded as one of the most effective tools for monitoring the retinal structures and their pathological changes [1]. Although OCT imaging has been improved markedly in both hardware and software, images still suffer from speckle noise brought by its intrinsic imaging mode and other motion artifacts during OCT inspect [1]. Image noise may obscure subtle but important structural details which are helpful to clinical diagnosis. Therefore, OCT image denoising has been extensively studied for improving visual examination and the performance of automatic analysis.

Various OCT image denoising algorithms have been developed based on traditional computer vision theories, including spatial-based (e.g. nonlinear diffusion filtering [2]) and transform-based methods (e.g. wavelet transform [3]). To

© Springer Nature Switzerland AG 2019
H. Fu et al. (Eds.): OMIA 2019, LNCS 11855, pp. 9–17, 2019.
https://doi.org/10.1007/978-3-030-32956-3_2

achieve better edge-preserving performance, several methods apply local image representation (e.g. image patches) for noise reduction [4,5]. The non-local means (NLM) performs weighted averaging within image patches extracted by a predefined searching window [4], and block-matching and 3D filtering (BM3D) groups similar patches into 3D format and performs collaborative filtering on each group to achieve the speckle noise reduction [5]. However, most existing methods require parameter tuning to match different noise distributions in OCT images [6], which restricts their usage in practice.

Recently, image denoising methods based on deep convolutional neural networks (CNN) have been developed, which eventually make it possible to achieve OCT noise reduction in an end-to-end manner [7,8]. A custom network, with the advantage of U-Net, residual learning and dilated convolution, is proposed to denoise single frame OCT B-scans [7]. Based on conditional generative adversarial networks (cGAN) [9], an edge-sensitive cGAN is proposed to noise reduction for OCT images and obtains remarkable denoising performance [8]. However, these CNN denoising frameworks need paired OCT images (noisy images and corresponding clean images) for training. The more commonly used method is to obtain clean images through registration and averaging of multi-frame B-scans [8]. In addition to requiring longer scanning time, this approach has two potential drawbacks. Firstly, field of interest cannot be absolutely aligned due to voluntary or involuntary eye movement during multiple scans. Secondly, some small pathological area (e.g. hyper reflection) could be blurred or totally removed after averaging. Instead, some methods synthesize noisy images with additive noise assumption (e.g. Gaussian noise in [7]), which does not make sense, because most OCT images are suffered from speckle noise or mixed noise sources.

In this paper, we propose a structure-aware noise reduction GAN (SNR-GAN) for unpaired OCT image denoising. The established adversarial network, inspired by cycle GAN [10], learns the mapping function between two domains (the clean and the noisy) rather than pixel-wise correspondence between paired images. We train the network with unpaired OCT images, which significantly reduce the difficulty of collecting training samples. Furthermore, we add an extra structural loss to preserve the fine structure of retinal tissues. Finally, the denoising performance is evaluated by visual inspection and quantitative metrics.

2 Proposed Method

In this section, we introduce the proposed network architecture, including the custom objective function with adversarial loss, cycle consistence loss and structure-aware loss. An overview of the proposed framework is illustrated in Fig. 1.

2.1 Design and Network Architecture

GAN can be considered as minimization-maximization game between generator and discriminator [11]. A generator creates new samples to maximize the probability of fooling the discriminator, while discriminator is trained to identify if

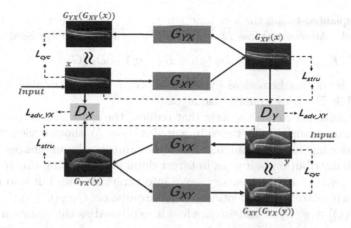

Fig. 1. Model architecture and loss function of our proposed SNR-GAN for unpaired OCT images (x, y). G_{XY} and G_{YX} are generators, D_X and D_Y are discriminators. L_{adv_XY}, L_{adv_YX}, L_{cyc}, and L_{stru} are the subcomponents of objective function.

the samples are real or fake. Cycle GAN has been proved as an effective way for unpaired image translation [10]. Different from traditional GANs, which only have one set of generator and discriminator, cycle GAN is composed of two sets of generators and discriminators to guarantee cycle-consistent translation between unpaired cross-domain images. Motivated by cycle GAN, we consider OCT noise reduction as image translation between clean and noisy image domains.

Suppose we have two OCT image domains, clean domain $x_i \in X$ and noisy do-main $y_i \in Y$. The proposed SNR-GAN is composed of two generators, G_{XY} and G_{YX}, and two discriminators D_X and D_Y. Firstly, two generators learn the image mappings between clean and noisy domain. G_{YX} translates the noisy OCT images to the clean image, while G_{XY} performs the inverse task. The network architecture for both generators is U-shape fully convolutional neural network, containing series of down and up sampling convolutional blocks. In this study, we used 8 convolution blocks for both down and up sampling, and each convolutional block has a stride of 2. Then, two discriminators are binary classifiers which distinguish the translated and the real images, in order to push generators to create realistic images. D_X calculates and compares the probability of x and $G_{YX}(y)$ being considered as clean images, and vice versa. Both discriminators are composed of 5 convolutional blocks.

2.2 Objective Function

The original objective function of cycle GAN includes two portions, adversarial and cycle consistence loss [12]. We introduce extra penalty based on structural similarity index (SSIM) [12] to preserve subtle features during noise reduction.

Adversarial loss can be subdivided into two parts for different mapping functions, $G_{XY}\colon X \mapsto Y$ and $G_{YX}\colon Y \mapsto X$. Here we only describe the loss function L_{adv_XY} for G_{XY} and the corresponding discriminator D_Y, because it can be easily transplanted to get the loss function L_{adv_YX} for G_{YX} and D_X following the same rule. Adversarial loss L_{adv_XY} for G_{XY} and D_Y is expressed as below:

$$L_{adv\_XY} = E_Y[\log D_Y(y)] + E_X[\log 1 - D_Y(G_{XY}(x))]$$

where $E[.]$ is the mathematical expectation. G_{XY} aims to minimize this loss function while D_Y tries to maximize it.

Cycle consistence loss is the item that reduces the space of possible mapping function during translation between unpaired images. Without cycle consistence loss, the mapping function can be trained to translate input images into any random sub-distribution of images in target domain. By adding this term, both generators need not only translate images into target domain but also translate them back which are almost identical to their inputs, i.e. $G_{YX}(G_{XY}(x)) \approx x$ and $G_{XY}(G_{YX}(y)) \approx y$. Cycle consistence loss is expressed as the equation below:

$$L_{cyc} = E_X[\|G_{YX}(G_{XY}(x)) - x\|_1] + E_Y[\|G_{XY}(G_{YX}(y)) - y\|_1]$$

where $E[.]$ is the mathematical expectation and $\|.\|_1$ is the L_1 norm.

In our application, we aim to remove noises instead of distorting or losing vital structural features. Even subtle structural features are important which may have influence on doctor diagnosis. SSIM is a commonly used indicator to assess structural similarity between two images [12], which is defined as follows:

$$SSIM = \frac{(2\mu_x + 2\mu_y + C_1)(2\sigma_{xy} + C_2)}{(\mu_x^2 + \mu_y^2 + C_1)(\sigma_x^2 + \sigma_y^2 + C_2)}$$

where μ_x and μ_y denote mean, σ_x and σ_y denote standard deviation, and σ_{xy} denotes covariance for two input images. C_1 and C_2 are constants. Based on SSIM calculation, we propose structure-aware loss to avoid geometrical distortion during image translation. SSIM loss is expressed as the equation below:

$$L_{ssim} = -E_X[SSIM(G_{XY}(x), x)] - E_Y[SSIM(G_{YX}(y), y)]$$

As defined above, SSIM loss performs the comparison between original image and corresponding translated image only through one generator, while cycle consistence loss compares the original image with its corresponding translated images created by concatenated operation of two generators.

SSIM loss mentioned above is calculated on entire image, but more attention is deserved in the critical regions during denoising. An ideal noise reducer tends to remove all noise on background while enhance the signal in foreground. We therefore introduce regional SSIM loss which calculates SSIM in the dedicated regions instead of the whole image. Regional SSIM loss is then defined as below:

$$L_{r-ssim} = -E_X[SSIM(sub(G_{XY}(x)), sub(x)))] - E_Y[SSIM(sub(G_{YX}(u)), sub(u)))]$$

where $sub(.)$ stands for sub-region, and its location is manually marked with bounding box on retinal area. Finally, the full objective function is defined as:

$$L_{total} = L_{adv} + \lambda_1 L_{cyc} + \lambda_2 L_{stru}$$
$$= L_{adv\_XY} + L_{adv\_YX} + \lambda_1 L_{cyc} + \lambda_2(L_{ssim} + 0.5 * L_{r-ssim})$$

where coefficients λ_1 and λ_2 are used to control the relative importance of three items. Each item plays different roles in creating high quality images. Adversarial loss L_{adv} enables the network to learn the different distributions of image domains and builds the mapping functions between clean domain and noisy domain. Cycle consistence loss L_{cyc} adds constrains during image translation so that the translated images are distributed evenly in target domain. Finally, structure-aware loss L_{stru} adds more structural constrains on the translation by comparing original and translated images directly.

3   Experiments and Results

In our experiments, the training and testing images were acquired using a Spectralis OCT device (Heidelberg Engineering, Germany). We compared denoising performance of our method with traditional methods including BM3D [5], NLM [4] and K-SVD [13]. The quantitative validation was performed on normal and pathological images, via the metrics of signal-to-noise ratio (SNR), contrast-to-noise ratio (CNR) and SSIM. We did not compare our method with existing CNN based denoising method, since strictly paired data is not available.

3.1   Implementation Details

We selected 250 clean images and 210 noisy images for training based on SNR of entire image. Images that have SNR larger than 30 dB are considered as clean, while less than 20 dB are considered as noisy. Data were exported from Spectralis system and saved as 8-bits depth images. The image resolution is 768*497. All images were padded to square and then re-sized to 512*512. The study was approved by the medical ethics committee of the hospital. No data augmentation was applied during training. Adam solver with initial learning rate 2e-4 was applied to optimize the adversarial networks. After trial and error, the coefficients λ_1 and λ_2, mentioned in objective function, were set to 10 and 5 empirically. We trained the network for 50 epochs with batch size of 1. The proposed network was implemented in Python with Tensorflow and trained by using 4 NVIDIA Tesla P100 GPUs.

Because there are no paired OCT images, we define the denoising metrics of SNR and CNR on designated signal and background regions. These regions of interest (ROIs) were manually marked on each image, as shown in Figs. 2(a) and 3(a), where red rectangles represent signal regions and blue rectangle represents background region. For normal images, three red rectangles locate in retinal neural fiber layer, inner retina and retinal pigment epithelium (Fig. 2(a)). For pathological images, three red rectangles locate in typical abnormal regions such as hyperreflective foci and subretinal fluid (Fig. 3(a)). SNR reflects the level of noise in an image, which is defined as:

$$SNR = 10\log_{10}(\frac{max(I)^2}{\sigma_b^2})$$

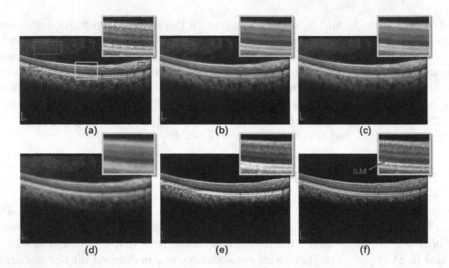

Fig. 2. Denoising results in a normal image. Original noisy image with selected ROIs for SNR and CNR calculation (a), and denoised images by BM3D (b), NLM (c), K-SVD (d), our methods without L_{stru}(e) and with L_{stru}(f). Region with yellow rectangle is zoomed. External limiting membranes (ELM) is marked for visual inspection. (Color figure online)

where $max(I)$ denotes the maximum grayscale in signal region and σ_b denotes standard deviation of the background region on the same image. CNR stands for the contrast between signal and background, which can be calculated as:

$$CNR_i = 10\log_{10}(\frac{|\mu_i - \mu_b|}{\sqrt{\sigma_i^2 - \sigma_b^2}})$$

where μ_i, μ_b and σ_i, σ_b denote mean and standard deviation for i-th signal region and background region, respectively. We averaged SNR and CNR over three selected signal regions. And we applied SSIM to evaluate structural fidelity.

3.2   Results

The validation was performed on 50 normal OCT images and 50 pathological OCT images with varying degrees of noise. We compared the denoising performance among three traditional denoising methods and our proposed methods. Figures 2 and 3 show the denoising results in normal and pathological datasets, respectively. And Table 1 shows the quantitative comparison among different denoising methods in both datasets.

By visual inspection, all methods suppressed the noises to some extend in background and retinal areas, while enhanced the contrast between retinal layers. These subjective impressions were confirmed by the metrics of SNR and CNR. All denoising methods had higher SNR and CNR compared to the original images (Table 1), while the highest SNR and CNR was achieved by the

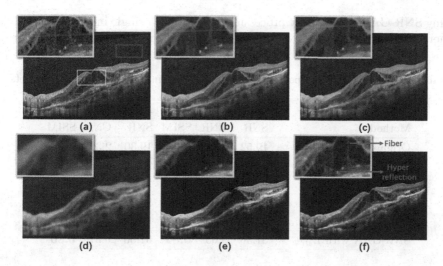

Fig. 3. Denoising results in a typical pathological image. Original noisy image with selected ROIs for SNR and CNR calculation (a); and denoised images with BM3D (b), NLM (c), K-SVD (d), our methods without $L_{stru}(e)$ and with $L_{stru}(f)$. Region with yellow rectangle is zoomed. Fiber and hyper reflection are marked for visual inspection. (Color figure online)

proposed SNR-GAN without L_{stru}. Higher SNR indicates relatively higher level of signal compared with noise level, and higher CNR means better contrast for image observation. Our proposed method cleaned up most of speckle noise on background region (Fig. 2(e–f), Fig. 3(e–f)), while BM3D, NLM and K-SVD were only capable of removing part of noise, that actually created inhomogeneous background in both datasets (Fig. 2(b–d), Fig. 3(b–d)).

Structural fidelity is another metric for noise reduction, especially for the fine retinal structures. In Figs. 2 and 3, it was obvious that BM3D, NLM and K-SVD tended to over smoothing noisy images which destroyed the textures. In normal images, the external limiting membrane layer (marked in Fig. 2(f)) was retained after denoising by our SNR-GAN with L_{stru}. Meanwhile, in pathological images, there are still some fibrous connective tissues (marked in Fig. 3(f)) within subretinal fluid, which is very important to evaluate the stage of disease during clinical diagnosis. Our SNR-GAN with L_{stru} preserved these subtle tissues while reducing image noise effectively, intuitively shown by the highest SSIM index in Table 1 and visual inspection, while original cycle GAN tended to remove subtle fibrous connective tissues, shown in Fig. 3(e).

In general, our SNR-GAN with L_{stru} achieved the best denoising performance while considering noise reduction as well as fine structural preservation. There is a slight decay in SNR and CNR, compared to the performance of original cycle GAN, mainly because minor noise in background is reserved by adding structural constraints. Besides, our method has the advantage in time efficiency.

Using SNR-GAN, the average processing time for each single image is 0.3 s, while other methods take 30 s or even longer under the same test environment.

Table 1. Quantitative comparison of different denoising methods in two datasets.

Method	Normal dataset			Pathological dataset		
	SNR	CNR	SSIM	SNR	CNR	SSIM
Original	19.52	5.31	N/A	19.30	2.32	N/A
BM3D	28.62	8.22	0.52	28.09	5.88	0.40
NLM	23.96	6.87	0.79	27.72	6.07	0.54
K-SVD	20.34	5.82	0.48	24.74	3.85	0.56
SNR-GAN without L_{stru}	**38.31**	**9.44**	0.65	**35.10**	**6.97**	0.61
SNR-GAN with L_{stru}	31.36	9.05	**0.84**	31.50	6.31	**0.80**

4 Conclusion

In this paper, we introduced an end-to-end deep network for OCT image denoising. Our method was motivated by cycle GAN, but denoising performance was further improved by introducing global and regional SSIM loss. Compared to other methods, we achieved better noise reduction performance as well as fine structural preservation in both normal and pathological OCT datasets. The greatest advantage of our proposed method lies in that it does not re-quire paired training samples. Related denoising works done by other researchers, using cGAN [8] or other convolutional neural network [7], always require paired training samples. Paired samples could not be obtained easily and strictly paired data is not available. In fact, there are ways to get large amount of un-paired clean and noisy OCT images. Our proposed method is a novel and quite practical framework to collect training data and obtain high-quality denoising model, which can be used for OCT images and even other medical image modalities.

References

1. Adhi, N., Duker, J.S.: Optical coherence tomography-current and future applications. Curr. Opin. Ophthalmol. **24**, 213–221 (2013)
2. Salinas, H.M., Fernández, D.C.: Comparison of PDE-Based nonlinear diffusion approaches for image enhancement and denoising in optical coherence tomography. IEEE Trans. Med. Imaging **26**(6), 761–771 (2007)
3. Mayer, M.A., Borsdorf, A., Wagner, M., et al.: Wavelet denoising of multiframe optical coherence tomography data. Biomed. Opt. Express **3**(3), 572–589 (2012)
4. Aum, J., Kim, J.H., Jeong, J.: Effective speckle noise suppression in optical coherence tomography images using nonlocal means denoising filter with double Gaussian aniso-tropic kernels. Appl. Opt. **54**(13), D43–D50 (2015)

5. Chong, B., Zhu, Y.: Speckle reduction in optical coherence tomography images of human finger skin by wavelet modified BM3D filter. Opt. Commun. **291**, 461–469 (2013)
6. Li, M., Idoughi, R., Choudhury, B., et al.: Statistical model for OCT image denoising. Biomed. Opt. Express **8**(9), 3903–3917 (2017)
7. Devalla, S.K., Subramanian, G., Pham, T.H., et al.: A deep learning approach to denoise optical coherence tomography images of the optic nerve head. arXiv preprint arXiv:1809.10589 (2018)
8. Ma, Y., Chen, X., Zhu, W., et al.: Speckle noise reduction in optical coherence tomography images based on edge-sensitive cGAN. Biomed. Opt. Express **9**(11), 5129–5146 (2018)
9. Isola, P., Zhu, J.Y., Zhou, T., et al.: Image-to-image translation with conditional adversarial networks. In: Computer Vision and Pattern Recognition (2017)
10. Zhu, J.Y., Park, T., Isola, P., et al.: Unpaired image-to-image translation using cycle-consistent adversarial networks. In: International Conference on Computer Vision (2017)
11. Goodfellow, I., Pouget-Abadie, J., Mirza, M., et al.: Generative adversarial networks. In: Advances in Neural Information Processing Systems, Montreal (2014)
12. Wang, Z., Bovik, A.C., Sheikh, H.R., et al.: Image quality assessment: from error visibility to structural similarity. IEEE Trans. Image Process. **13**(4), 600–612 (2004)
13. Kafieh, R., Rabbani, H., Selesnick, I.: Three dimensional data-driven multi scale atomic representation of optical coherence tomography. IEEE Trans. Med. Imag. **34**(5), 1042–1062 (2015)

Region-Based Segmentation of Capillary Density in Optical Coherence Tomography Angiography

Wenxiang Deng[1,2], Michelle R. Tamplin[2,3], Isabella M. Grumbach[2,3], Randy H. Kardon[2,4], and Mona K. Garvin[1,2(✉)]

[1] Department of Electrical and Computer Engineering, The University of Iowa, Iowa City, IA, USA
mona-garvin@uiowa.edu
[2] Iowa City VA Health Care System, Iowa City, IA, USA
[3] Department of Internal Medicine, The University of Iowa, Iowa City, IA, USA
[4] Department of Ophthalmology and Visual Sciences, The University of Iowa, Iowa City, IA, USA

Abstract. Microvascular changes are one of the early symptoms of retinal diseases. Recently developed optical coherence tomography angiography (OCTA) technology allows visualization and analysis of the retinal microvascular network in a non-invasive way. However, automated analysis of microvascular changes in OCTA is not a trivial task. Current approaches often attempt to directly segment the microvasculature. These approaches generally have problems in cases of poor image quality and limited visibility of the vasculature. Evaluating the quality of the results is also challenging because of the difficulty of manually tracing the microvasculature, especially in cases of low image quality or with images with a larger field of view. In this work, we develop an automated deep-learning approach to assign each pixel within human OCTA *en-face* images the probability of belonging to a microvascular density region of each of the following categories: avascular, hypovascular, and capillary-dense. The AUCs (area under the receiver operating characteristic curves) were 0.99 (avascular), 0.93 (hypovascular), and 0.97 (capillary-dense) for segmenting each of the categories. The results show very good performance and enables global and region-based quantitative estimates of microvascular density even in relatively low-quality *en-face* images.

Keywords: OCT-Angiography · Capillary density · Deep learning

1 Introduction

The retinal microvascular network supplies the retina with oxygen and nutrition. Ocular diseases including diabetic retinopathy [1], glaucoma [2], age-related

H. Fu et al. (Eds.): OMIA 2019, LNCS 11855, pp. 18–25, 2019.
https://doi.org/10.1007/978-3-030-32956-3_3

macular degeneration [3], and radiation retinopathy [4,5] often show changes in the microvasculature. Traditionally, fluorescein angiography (FA) is widely used for assessing retinal micro-circulation. However, this technique is invasive and doesn't distinguish microvasculature in superficial retinal layers from the deeper plexus. OCT angiography (OCTA) is a modality increasingly used for assessing retinal vasculature pathologies in retinal studies [6–8].

Many existing approach for measuring capillary density from *en-face* OCTA images – including the use of fractal analysis [9,10], vessel enhancement filters [11], and a convolutional-neural-network approach [12] – attempt to first directly segment the microvasculature. However, because the quality of OCTA images varies and is susceptible to noise or artifacts, a direct measurement of vessels in OCTA *en-face* images is often not feasible. Figure 1 shows an example comparison of good quality *en-face* images (Fig. 1(a), (b)) and an image with a poorly visible microvascular network (Fig. 1(c)). These differences can be common in the analysis of patients with diseases such as radiation retinopathy [4].

Other strategies for providing an overall sense of vessel density from an OCTA *en-face* image include directly detecting the fovea avascular zone (FAZ) or non-FAZ avascular regions using filters and/or fractal analysis [13–15] and a recent deep-learning approach to detect avascular areas [16]. However, these approaches do not take into consideration of the possibility of reduced (but not completely avascular) capillary density in regions. Also, location-dependent artifacts are often not addressed.

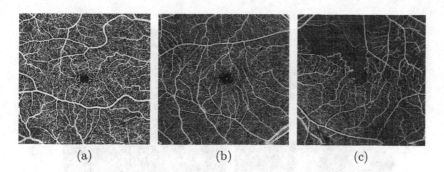

(a)   (b)   (c)

Fig. 1. Example variability of OCTA *en-face* image quality. (a) An example image from a healthy subject showing vessels clearly. (b) and (c) are *en-face* images from the unaffected and affected eye of a patient with radiation retinopathy. (c) shows an image with a lower signal-to-noise ratio (SNR) and less visible capillaries.

In this work, to address the frequent limited visibility of individual capillaries in OCTA images but to also allow for characterization of differing density levels at each pixel, we train and evaluate a deep-learning-based approach (using a modified U-net architecture) to segment avascular, hypovascular, and capillary-dense areas from OCTA *en-face* images. We train and evaluate our approach (using 10-fold cross-validation) on a challenging dataset of uveal melanoma patients (including those with radiation retinopathy) and controls whereby, as shown in Fig. 1, the visibility of the capillaries can vary substantially.

2 Methods

2.1 Preprocessing

In each OCTA scan we use in this work, there is a volumetric structural OCT scan showing the macular tissue and a corresponding split spectrum amplitude-decorrelation angiography (SSADA) scan showing the vesselness. We use the raw OCT and SSADA data to segment and form the *en-face* projection angiogram of superficial layers. More specifically, to obtain the layer information, a graph-based layer segmentation algorithm [17] is used to segment the superficial layers in the corresponding OCT volumes. In this approach, we follow the definition from Camino et al. [18] and use the superficial layers as the combination of retinal nerve fiber layer (RNFL) to inner plexiform layer (IPL). Figure 2(a) shows the SSADA information overlaid on a b-scan in the segmented layers. To generate the projection images, we use the maximum intensity in each column in these layers. An example result can be seen in Fig. 2(b).

Because the deep learning approach we use prefers pixel dimensions divisible by 32, the generated *en-face* images are also unified to dimensions of 480 × 480 by upsampling with cubic spline interpolation to preserve more details.

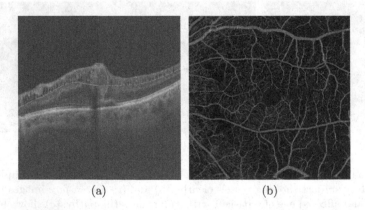

(a)   (b)

Fig. 2. An example OCTA image. (a) OCT b-scan showing the superficial layers used in this study. The superficial layers are shown between the red and green boundaries, and only the SSADA information shown in red scatters between the boundaries are used. (b) The *en-face* projection angiogram of the superficial layers of the structural OCT volume indicated in (a). (Color figure online)

2.2 Input Data Generation

For the input of our deep network, in addition to the *en-face* images, to help the network deal with potential location-specific artifacts, such as signal dropout on the temporal side (Fig. 3(a)), we also provide a radial location map (providing the distance from the center), as shown in Fig. 3(c), and a temporal-to-nasal location map (providing the distance from the temporal side), as shown in (Fig. 3(d)). By providing such additional inputs, we can help to avoid results such as that in Fig. 3(b) whereby the regions with signal dropout would incorrectly be classified as avascular or hypovascular regions.

In the training stage, rotation, flipping, and cropping are applied to each input image before each training iteration. The original *en-face* image is first is augmented with random rotations in range $(-180°, 180°)$. A random horizontal flip is then applied. The training image is then randomly cropped to a fixed size of 256×256 pixels and, to further help address the possibility of regional signal dropout, a local contrast change is also applied.

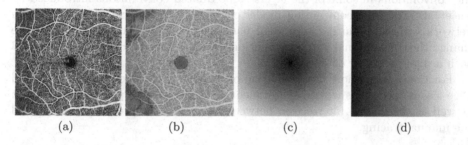

(a)   (b)   (c)   (d)

Fig. 3. Motivation for use of location maps as additional input. (a) OCTA *en-face* image from a healthy control subject with local signal dropout on temporal side. (b) Segmentation result without additional location-specific maps as input. (c) Radial location map. (d) Temporal-to-nasal location map.

2.3 Network Structure

Here, we develop a fully automated deep-learning based approach. A modified version of U-Net [19] is applied to do a pixel-based separation of avascular regions, hypovascular regions (i.e., regions with capillary dropout) and capillary-dense regions. The overall network structure of this work is shown in Fig. 4. The inputs are the *en-face* image and two location maps as discussed above and the outputs are three probability maps for the avascular region, hypovascular region, and capillary-dense region.

Our modifications to the original U-net are, in part, inspired by recent work by Iglovikov et al. [20] that showed that using a deeper VGG11 network as the encoder can increase the segmentation accuracy compared to the original U-Net. Here we use a similar network structure. We also add batch normalization (BN)

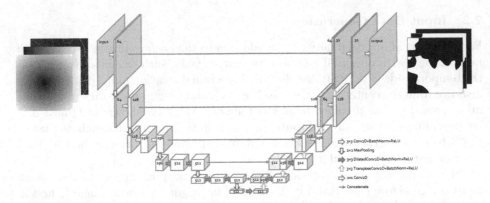

Fig. 4. The modified version of U-Net with dilated convolutions used in this work.

layers [21] to improve the numerical stability. Another modification is changing the convolutions in some of the layers to be dilated convolutions introduced by Yu et al. [22]. This convolution model enables a larger receptive field for the network without requiring additional parameters to be learned. Therefore, we change the conventional 3×3 convolutions in some layers of the U-Net to 3×3 with a dilation factor of 2.

For the loss function, a combination of binary cross-entropy (BCE) loss and Dice loss is used, thus we can simultaneously maximize the per-pixel prediction as well as the overall intersection between the predicted probability maps and the manual tracing.

3   Experiments and Results

A total of 166 macular OCTA scan sets (structural OCT volumes and corresponding SSADA data) acquired with AngioVue (Optovue, Inc., Fremont, CA) from 59 human subjects are used in the study. Within the subjects, 43 are uveal melanoma patients and 16 are control cases. All procedures involving human subjects in this study were approved by the Institutional Review Boards (IRB) and the Human Subjects Office at the University of Iowa.

The OCT volumes and SSADA scans from all control subjects and 37/43 uveal melanoma subjects have dimensions of $400 \times 400 \times 640$ and $400 \times 400 \times 160$, respectively, corresponding to physical dimensions of $6\,mm \times 6\,mm \times 2\,mm$. The OCT volumes and SSADA scans from the remaining six uveal melanoma subjects have dimensions of $304 \times 304 \times 640$ and $304 \times 304 \times 160$, respectively. As mentioned in Sect. 2.1, all the generated *en-face* images are rescaled to a unified size of 480×480 before further processing. For training and evaluation purposes, the capillary-dense regions, hypovascular (capillary dropout) and avascular regions are also manually identified in each *en-face* image.

To train the network, we use Adam optimizer to train the network from scratch, with an initial learning rate of 1e-4. The learning rate changed to 1e-5

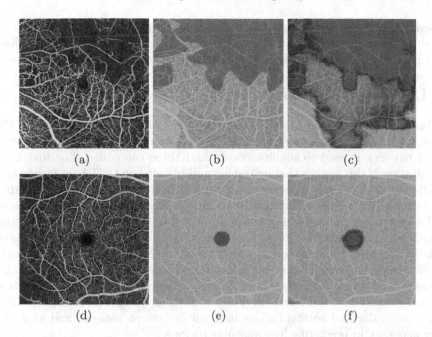

Fig. 5. Example results from the proposed deep-learning approach. (a–c) show an example result from a radiation retinopathy patient, and (d–f) are for a control subject. (a), (d) The *en-face* OCTA image (enhanced for increased visibility). (b), (e) The manual tracing for the image. (c), (f) The segmentation result of the three classes. Red = avascular. Orange = hypovascular. Green = capillary-dense. (Color figure online)

after 500 epochs and 1e-6 at 700 epochs. A total of 1000 epochs are used for the total training on a single Nvidia GeForce 1080 Ti GPU. It takes around 3 hours to train one network, and approximately a day to train all the U-Nets for the cross-validation experiment. In the testing phase, for each 480×480 *en-face* image, it takes about 0.1 s to run on GPU and 3 s to run on a CPU. Because the U-net structure is fully convolutional, while training is performed with 256×256 cropped images, testing can be directly performed with 480×480 images.

To evaluate the results, area under the pixel-based receiver operating characteristic (ROC) curves (AUC) are used in a 10-fold cross-validation approach. The first nine folds each have images from 53 randomly chosen subjects as the training set and images from 6 subjects as the test set. The remaining fold has 54 subjects for training and 5 for testing. Among the ROCs, the prediction of avascular regions has the best performance, with an AUC of 0.99 and the prediction of the capillary-dense regions has the second-best performance, with an AUC of 0.97. The hypovascular prediction generally gives a slightly lower AUC of 0.93. This is partially because the tracing is more subjective than the tracing of avascular areas as it is sometimes hard to distinguish hypovascular regions.

An example of the original image, the manual tracing, and the corresponding result can be seen in Fig. 5. Visually, this approach offers clean predictions for avascular regions marked in red. In fact, the regions are slightly more consistent and detailed than manual tracings. On the other hand, the predicted hypovascular regions in orange are less confident and have some misclassified regions.

4   Discussion and Conclusions

In this work, we develop a region-based segmentation method to identify different microvascular densities within OCTA *en-face* projection images. A deep neural network is used to simultaneously find these categories. The first main contribution of this work is that by tracing and training on OCTA *en-face* images as different density regions, we can avoid the difficulty to distinguish individual capillaries giving us the ability to still obtain density estimates on lower-quality OCTA images where individual vessels are not visible. The second contribution is that by adding additional location maps to the inputs of the deep learning network, the segmentation network can successfully avoid some OCTA artifacts to generate better results. To the best of our knowledge, this trained deep-learning approach is the first fully automated approach to categorize multiple levels of vascular density in OCTA images. The results show very accurate predictions for the avascular and normal regions in uveal melanoma patients and a slightly lower accuracy in predicting hypovascular regions.

Acknowledgements. This work was supported, in part, by the VA Center for the Prevention and Treatment of Visual Loss, AHA 18IPA34170003, and NIH T32 CA078586.

References

1. Chen, Q., et al.: Macular vascular fractal dimension in the deep capillary layer as an early indicator of microvascular loss for retinopathy in type 2 diabetic patients. Invest. Ophthalmol. Vis. Sci. **58**(9), 3785–3794 (2017)
2. Flammer, J., et al.: The impact of ocular blood flow in glaucoma. Prog. Retinal Eye Res. **21**(4), 359–393 (2002)
3. Jia, Y., et al.: Quantitative optical coherence tomography angiography of choroidal neovascularization in age-related macular degeneration. Ophthalmology **121**(7), 1435–1444 (2014)
4. Veverka, K.K., AbouChehade, J.E., Iezzi Jr., R., Pulido, J.S.: Noninvasive grading of radiation retinopathy: the use of optical coherence tomography angiography. Retina **35**(11), 2400–2410 (2015)
5. Shields, C.L., Say, E.A.T., Samara, W.A., Khoo, C.T., Mashayekhi, A., Shields, J.A.: Optical coherence tomography angiography of the macula after plaque radiotherapy of choroidal melanoma: comparison of irradiated versus nonirradiated eyes in 65 patients. Retina **36**(8), 1493–1505 (2016)
6. Spaide, R.F., Klancnik, J.M., Cooney, M.J.: Retinal vascular layers imaged by fluorescein angiography and optical coherence tomography angiography. JAMA Ophthalmol. **133**(1), 45–50 (2015)
7. Giannakaki-Zimmermann, H., Kokona, D., Wolf, S., Ebneter, A., Zinkernagel, M.S.: Optical coherence tomography angiography in mice: comparison with confocal scanning laser microscopy and fluorescein angiography. Transl. Vis. Sci. Technol. **5**(4), 11–11 (2016)

8. Chen, J.J., AbouChehade, J.E., Iezzi Jr., R., Leavitt, J.A., Kardon, R.II.. Optical coherence angiographic demonstration of retinal changes from chronic optic neuropathies. Neuro-Ophthalmology **41**(2), 76–83 (2017). https://doi.org/10.1080/01658107.2016.1275703
9. Gadde, S.G., et al.: Quantification of vessel density in retinal optical coherence tomography angiography images using local fractal dimension. Invest. Ophthalmol. Vis. Sci. **57**(1), 246–252 (2016)
10. Zahid, S., et al.: Fractal dimensional analysis of optical coherence tomography angiography in eyes with diabetic retinopathy. Invest. Ophthalmol. Vis. Sci. **57**(11), 4940–4947 (2016)
11. Dongye, C., et al.: Automated detection of dilated capillaries on optical coherence tomography angiography. Biomed. Opt. Express **8**(2), 1101–1109 (2017)
12. Prentašić, P., et al.: Segmentation of the foveal microvasculature using deep learning networks. J. Biomed. Opt. **21**(7), 075008–075008 (2016)
13. Zhang, M., Hwang, T.S., Dongye, C., Wilson, D.J., Huang, D., Jia, Y.: Automated quantification of nonperfusion in three retinal plexuses using projection-resolved optical coherence tomography angiography in diabetic retinopathy. Invest. Ophthalmol. Vis. Sci. **57**(13), 5101–5106 (2016)
14. Sandhu, H.S., et al.: Automated diabetic retinopathy detection using optical coherence tomography angiography: a pilot study. Br. J. Ophthalmol. **102**(11), 1564–1569 (2018)
15. Anegondi, N., Chidambara, L., Bhanushali, D., Gadde, S.G., Yadav, N.K., Sinha Roy, A.: An automated framework to quantify areas of regional ischemia in retinal vascular diseases with OCT angiography. J. Biophotonics **11**(2), e201600312 (2018)
16. Guo, Y., Camino, A., Wang, J., Huang, D., Hwang, T.S., Jia, Y.: MEDnet, a neural network for automated detection of avascular area in OCT angiography. Biomed. Opt. Express **9**(11), 5147–5158 (2018)
17. Garvin, M.K., Abràmoff, M.D., Wu, X., Russell, S.R., Burns, T.L., Sonka, M.: Automated 3-D intraretinal layer segmentation of macular spectral-domain optical coherence tomography images. IEEE Trans. Med. Imaging **28**(9), 1436–1447 (2009)
18. Camino, A., et al.: Automated registration and enhanced processing of clinical optical coherence tomography angiography. Quant. Imaging Med. Surg. **6**(4), 391 (2016)
19. Ronneberger, O., Fischer, P., Brox, T.: U-Net: convolutional networks for biomedical image segmentation. In: Navab, N., Hornegger, J., Wells, W.M., Frangi, A.F. (eds.) MICCAI 2015. LNCS, vol. 9351, pp. 234–241. Springer, Cham (2015). https://doi.org/10.1007/978-3-319-24574-4_28
20. Iglovikov, V., Shvets, A.: TernausNet: U-Net with VGG11 encoder pre-trained on ImageNet for image segmentation. arXiv preprint arXiv:1801.05746 (2018)
21. Ioffe, S., Szegedy, C.: Batch normalization: accelerating deep network training by reducing internal covariate shift. In: International Conference on Machine Learning, pp. 448–456 (2015)
22. Yu, F., Koltun, V.: Multi-scale context aggregation by dilated convolutions. arXiv preprint arXiv:1511.07122 (2015)

An Amplified-Target Loss Approach for Photoreceptor Layer Segmentation in Pathological OCT Scans

José Ignacio Orlando[1(✉)], Anna Breger[2], Hrvoje Bogunović[1], Sophie Riedl[1], Bianca S. Gerendas[1], Martin Ehler[2], and Ursula Schmidt-Erfurth[1]

[1] Christian Doppler Laboratory for Ophthalmic Image Analysis (OPTIMA), Department of Ophthalmology and Optometry, Medical University of Vienna, Vienna, Austria
jose.orlando@meduniwien.ac.at
[2] Department of Mathematics, University of Vienna, Vienna, Austria

Abstract. Segmenting anatomical structures such as the photoreceptor layer in retinal optical coherence tomography (OCT) scans is challenging in pathological scenarios. Supervised deep learning models trained with standard loss functions are usually able to characterize only the most common disease appearance from a training set, resulting in suboptimal performance and poor generalization when dealing with unseen lesions. In this paper we propose to overcome this limitation by means of an augmented target loss function framework. We introduce a novel amplified-target loss that explicitly penalizes errors within the central area of the input images, based on the observation that most of the challenging disease appearance is usually located in this area. We experimentally validated our approach using a data set with OCT scans of patients with macular diseases. We observe increased performance compared to the models that use only the standard losses. Our proposed loss function strongly supports the segmentation model to better distinguish photoreceptors in highly pathological scenarios.

1 Introduction

Supervised deep learning techniques have revolutionized the field of medical image segmentation [4], particularlly with fully convolutional neural network architectures such as the U-Net [6]. To learn these networks, a loss function L is optimized using gradient based approaches and backpropagation. This function is usually defined in terms of metrics that quantify the discrepancies between a trustworthy/ground truth labelling and the predicted segmentation.

In this typical framework a loss function is not explicitly tailored to aim for a specific feature in the target space. Hence, the network firstly learns the dominating characteristics of the target images in the training set, and its remaining

---

J. I. Orlando and A. Breger—Equal contribution.

© Springer Nature Switzerland AG 2019
H. Fu et al. (Eds.): OMIA 2019, LNCS 11855, pp. 26–34, 2019.
https://doi.org/10.1007/978-3-030-32956-3_4

capacity is gradually devoted to characterize other less prevalent target features. This becomes an issue when dealing with highly pathological data, where lesions or disease appearance might significantly differ between patients. To overcome this limitation, some authors proposed to train segmentation models using a linear combination of different losses such as cross-entropy and Dice [3]. However, these metrics are still computed from the same target representation, so they do not enhance a specific target feature. In this paper we propose to extend this idea by using the framework of *augmented target loss functions*, introduced in [1]. Rather than relying on a single or a linear combination of loss functions defined on the original prediction and target space, Breger *et al.* [1] proposed to compute the loss on alternative representations of the predictions and targets, obtained by applying differentiable transformations T that enhance specific characteristics.

This paper focuses on the application of an augmented target loss function for photoreceptor layer segmentation in retinal optical coherence tomography (OCT) scans of patients with macular diseases. OCT is the state-of-the-art technique for imaging the retina, as it brings volumetric information through a stack of 2D scans (B-scans) at a micrometric resolution [7]. Ophthalmic disorders such as diabetic macular edema (DME), retinal vein occlusion (RVO) and age-related macular degeneration (AMD) gradually affect photoreceptors while progressing. The abnormal accumulation of fluid due to these diseases significantly alters the retina, eventually leading to photoreceptor cell death. This last characteristic can be noticed through OCT imaging: first as a pathological thinning of the photoreceptor layer, and more lately as complete disruptions on it (Fig. 1, right). It has been observed that these abnormalities are highly correlated with focal vision impairment [8] and visual accuity loss when located at the central area of the retina [2]. Hence, the automated characterization of the morphology of the photoreceptor layer is relevant for efficient quantification of functional loss.

In this paper we build on top of the architectural innovations proposed in [5] by training such a model using an augmented target loss function. Fitting the framework we introduce a novel amplified-target loss that induces further penalization to errors within the central area of the B-scans. As the most challenging pathologies are usually observed at the central area of fovea-centered OCT scans, our hypothesis is that incorporating this loss function as a kind of regularizer enforces the network to better characterize disease appearance. We validate our approach using a series of OCT scans of patients with AMD, DME and RVO. Our results empirically show that the proposed loss functions improve the performance within the central millimeters of the retina compared to using traditional losses without compromising the performance in the entire volume.

2   Methods

2.1   Augmented Target Loss Functions for Image Segmentation

In a supervised learning problem we aim to learn a function f with $f_\theta(x) \approx y$, where θ denotes the free parameters and $S = \{(x,y)^{(i)}\}, 1 < i < N$ is a given training set with pairs of inputs x and ground truth labels y. In the context

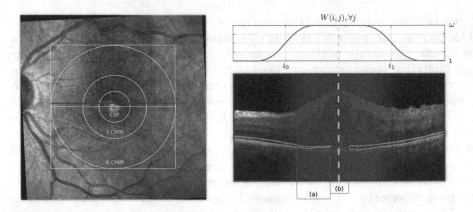

Fig. 1. Left: scanning laser ophthalmoscopy (SLO) of a patient with RVO. The square indicates the area captured by the OCT volume and the rings represent the central subfield (CSF) and the 3 and 6 central millimeters (3 CMM and 6 CMM). The blue line highlights the B-scan showed in the right side. Right: CSF B-scan with photoreceptor layer annotation (green) with (a) disruptions and (b) abnormal thinning. The red heat map represents the weighting strategy applied in our loss function. The central coordinate of the image is indicated with the yellow dotted line, and a profile of the weighting strategy is illustrated on top of the B-scan. (Color figure online)

of image segmentation, x corresponds to an input image, y and $\hat{y}$ are manual and predicted segmentations and f_θ is some segmentation model (e.g. a fully convolutional neural network such as the U-Net [6]).

To adjust the weights θ from the chosen network structure f_θ, a loss function L is minimized using gradient based optimization. L is a piecewise differentiable loss function, e.g. cross-entropy (CE) or mean square error (MSE), that measures the pixel-wise differences between $\hat{y}$ and y. In standard settings no specific areas of the images are penalized more than others. Thus, the parameters θ are mostly adjusted to characterize those features from the training set that have the most impact on the overall loss. Although this might be helpful to segment healthy anatomy, in pathological scenarios the network will overfit the prevalent features unless explicit regularization is imposed during training.

Here, we propose to use the framework of augmented target (AT) loss functions, introduced in [1]. These losses take into account prior knowledge of target characteristics via error estimation in transformed target spaces. The framework can be applied to any supervised learning problem based on loss optimization where additional information about the target data is available, provided it can be formulated as a transformation function T. The transformation may correspond to any piecewise differentiable function on the target space that yields a more beneficial representation of some known target characteristic.

Following [1], the AT loss functions L_{AT} is a linear combination of losses applied to transformed targets. Its general form is:

$$L_{AT} = \sum_{j=1}^{d} \lambda_j \cdot L^j \left(\{T_j(y_i)\}, \{T_j(\hat{y}_i)\} \right), \tag{1}$$

where $\lambda_j > 0$ corresponds to some weight, T_j to a specific transformation and L^j to some loss function, for all $j \in \{1, \ldots, d\}$.

Setting typically T_1 to the identity and L^1 to a standard loss, the additional terms in the L_{AT} loss act as amplified target information, yielding a new optimization problem:

$$\hat{\theta} = \arg\min_{\theta} \{ \lambda_1 \cdot L_1(\{y_i\}, \{\hat{y}_i\}) + \sum_{j=2}^{d} \lambda_j \cdot L^j \left(\{T_j(y_i)\}, \{T_j(\hat{y}_i)\} \right), \tag{2}$$

where the weights λ_1 and $\{\lambda_j\}_{j=2}^{d}$ control the balance between the main loss and the regularization terms respectively.

2.2 Amplified-Target Loss Functions for Photoreceptor Layer Segmentation

We experimentally study the AT loss function framework in the context of photoreceptor layer segmentation in pathological OCT scans. We tailor a so called *amplified-target loss* in which a transformation T is designed to bring an increased penalty to errors within the central area of the images. This loss is intended to incorporate the prior knowledge that abnormalities such as pathological thinnings and disruptions of the photoreceptor layer are more common in the central millimeters of the foveal area. To do so, we define a transformation $T(y_i) = \langle y_i, W \rangle$, where y_i corresponds to the given binary targets and W represents a weighting matrix that encodes a penalization weight for errors. This operation can analogously be applied to the predictions $\hat{y}_i$. Figure 1 graphically illustrates the design of the weighting matrix W. Formally, we define $W = G_\sigma * V$, where G_σ stands for a Gaussian filter with standard deviation σ. We define V as:

$$V_{i,j} := \begin{cases} \omega & \text{for } i_0 < i < i_1 \text{ and all } j, \\ 1 & \text{otherwise,} \end{cases} \tag{3}$$

where ω denotes the maximum weight assigned to the central area and $[i_0, i_1]$ is the horizontal interval of the image that is amplified. The Gaussian filter G_σ is used to smooth the penalization factor within the edges of the interval.

Following the formulation in (2), we can then redefine our empirical risk minimization problem as

$$\hat{\theta} = \arg\min_{\theta} \{ \lambda_1 \cdot L^1(\{y_i\}, \{\hat{y}_i\}) + \lambda_2 \cdot L^2 (\{\langle y_i, W \rangle, \langle \hat{y}_i, W \rangle\}) \}, \tag{4}$$

where we choose $\lambda_1, \lambda_2 \in \mathbb{R}$ and $L^1 = L^2$ as CE or MSE losses.

3   Experimental Setup

3.1   Materials

Our method was trained and tested on an in-house data set with 53 Spectralis OCT volumes of patients suffering from DME (10), RVO (27) and AMD (10). Each image comprises 496×512 pixels per B-scan, 49 B-scans per volume. All the B-scans were manually annotated by certified readers under the supervision of a retina expert, who modified the labels when necessary to obtain ground truth segmentations. The set was randomly divided into a training, a validation and a test set, each of them with 34, 4 and 15 scans, respectively, with approximately the same distribution of diseases and percentages of disrupted columns per B-scan (or A-scans).

3.2   Network Architecture and Training Setup

We used the photoreceptor segmentation network described in [5] in our experiments (note that any other architecture could be applied within our framework). We used as baselines CE and MSE comparing it to the adapted AT loss.

Every configuration was trained at a B-scan level with a batch size of 2 images, using Adam optimization and early stopping. Hence, training was stopped if the validation loss did not improve for the last 45 epochs. The learning rate was set to $\eta = 0.0001$, and divided by 2 if the validation loss was not improved during the last 15 epochs. Data augmentation was used in the form of random horizontal flippings. Binary segmentations were retrieved as in [5] by thresholding the softmax scores of the photoreceptors class using the Otsu algorithm.

4   Results and Discussion

We evaluated the performance for segmenting the photoreceptor layer using the volume-wise Dice index, at the CSF, the 3 CMM, the 3-1 ring and the full volume (Fig. 1, left). All the experiments with our AT loss functions were conducted using fixed values for $\sigma = \frac{1}{16}X$, $i_0 = \frac{1}{4}X$ and $i_1 = \frac{3}{4}X$ (with $X = 512$ being the horizontal size of the B-scans, in pixels), without optimizing them on the validation set. Different configurations for $\omega = 2^k, k \in \{1, ..., 5\}$ and λ_1 and $\lambda_2 \in \{0.001, 0.01, 0.1, 1, 2, 4, 8\}$ were analyzed, and the best configuration according to Dice index on the validation set was then fixed to allow a fair comparison on the test set. From this model selection step, we observed that $\omega = 8$, $\lambda_1 = 1$ and $\lambda_2 = 8$ reported the best performance for the AT loss with categorical cross-entropy (CE), and $\omega = 32$, $\lambda_1 = \lambda_2 = 1$ for the AT loss with mean square error (MSE).

Figure 2 depicts boxplots with the quantitative performance of each model on the test set, compared with their corresponding baselines trained only with CE and MSE, for each evaluation area. The mean and standard deviation values of

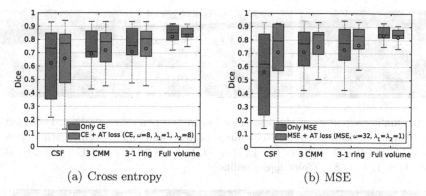

(a) Cross entropy (b) MSE

Fig. 2. Volume-wise Dice values for all the evaluated models and our proposed approach in each evaluation area. Circles indicate mean values. CSF: central subfield (1 central millimeter). 3 CMM: three central millimeter. 3-1 ring: area between CSF and 3 CMM.

Table 1. Volume-wise mean $\pm$ standard deviation Dice values in the test set for each photoreceptor segmentation model in each area.

Method	CSF	3 CMM	3-1 ring	Full volume
CE loss	0.622 ± 0.271	0.691 ± 0.242	0.708 ± 0.242	0.820 ± 0.118
CE + AT loss ($CE, \omega = 8, \lambda_1 = 1, \lambda_2 = 8$)	$\mathbf{0.656 \pm 0.256}$	$\mathbf{0.718 \pm 0.218}$	$\mathbf{0.732 \pm 0.218}$	$\mathbf{0.828 \pm 0.100}$
MSE loss	0.560 ± 0.303	0.707 ± 0.223	0.727 ± 0.223	$\mathbf{0.835 \pm 0.096}$
MSE + AT loss ($MSE, \omega = 32, \lambda_1 = \lambda_2 = 1$)	$\mathbf{0.708 \pm 0.254}$	$\mathbf{0.749 \pm 0.215}$	$\mathbf{0.760 \pm 0.213}$	0.821 ± 0.102

the Dice index are presented in Table 1. The incorporation of the AT loss allows to perform consistently better in all the cases, with the best results reported by the MSE loss. Statistical analysis using one-tail Wilcoxon sign-rank tests at a significance level $\alpha = 0.05$ showed that the model trained with MSE + AT loss reported significantly higher Dice values in the CSF area compared to using CE + AT loss or only MSE ($p < 0.0171$). These differences were not statistical significant with respect to the model trained with CE ($p = 0.1902$). When comparing the Dice values at the 3-1 ring, the MSE with AT loss model reported statistically significant better results than using only CE or MSE ($p < 0.0042$), which is consistent with its behavior in the 3 CMM ($p < 0.0416$). No statistically significant differences in performance were observed at the full volume level (two-tails test, $p > 0.0730$).

We qualitatively analyzed the segmentation and score maps using the CE and MSE combined with the AT loss. Figure 3 depicts segmentation results in a central B-scan from the test set, with score maps represented as heatmaps. Using MSE produces noisy scores within the lateral areas of the B-scans, and therefore spurious elements in the segmentation. CE, on the contrary, results in smoother score maps, although with few false negatives in the vicinity of subretinal fluid. This behavior is linked to the one observed in Table 1, where the MSE + AT

32 J. I. Orlando et al.

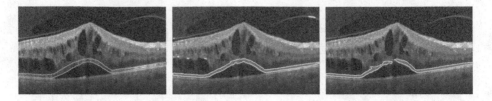

Fig. 3. Qualitative effect of the loss selection in the pixel score values. From left to right: manual annotation (green), score map (orange) and binary segmentation (yellow) obtained with MSE + AT loss (MSE, $\omega = 32, \lambda_1 = \lambda_2 = 1$) and CE + AT loss (CE, $\omega = 8$, $\lambda_1 = 1$, $\lambda_2 = 8$). (Color figure online)

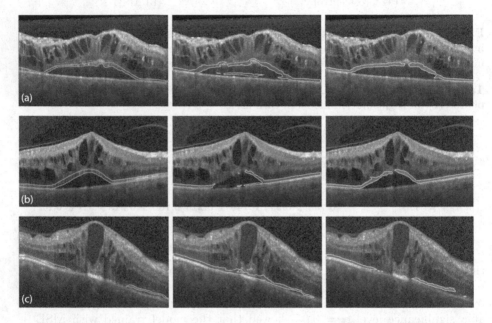

Fig. 4. Qualitative results in central B-scans from the test set. From left to right: manual annotations (green), results with only CE loss (blue) and results with CE + AT loss (CE, $\omega = 8$, $\lambda_1 = 1$, $\lambda_2 = 8$). (Color figure online)

loss model reported higher Dice in the central area than using CE, and smaller values in the full volume. The model trained with only MSE performs poorly in the CSF, the 3 CMM and the 3-1 ring, which indicate that it struggles to deal with pathologies. Similarly, the high performance at a volume level indicates that it can better characterize normal appearances. When using MSE + AT loss, a significant reduction in the amount of false negatives occurs at the central areas. However, as mentioned before, the score maps are noisy at the borders of the B-scans, which causes a drop in the full volume Dice. The model trained with CE + AT loss is less accurate at the center than the one trained with MSE + AT loss, but it still outperforms the baseline approaches. Moreover, at a volume basis the CE + AT loss remains competitive with respect to the one trained only with CE loss.

Finally, Fig. 4 presents qualitative results in exemplary central B-scans from our test set obtained both by the models trained with CE only and with CE + AT loss. Our approach produced more anatomically plausible segmentations than the standard CE loss in pathological areas with subretinal fluid (Fig. 4(a) and (b)) or large disruptions (Fig. 4(c)).

5 Conclusions

In this paper we proposed to use the framework of augmented target loss functions for photoreceptor layer segmentation in pathological OCT scans. We define an amplified-target loss incorporating a transformation that weights the central area of the input B-scans to further penalize errors commited in this region. We experimentally observed that this straightforward approach allows to significantly improve performance within the central millimeters of fovea-centered OCT scans, without affecting the overall performance in the entire volume. These results indicate that the proposed AT loss function acts as a form of regularization, better characterizing photoreceptors appearance within highly pathological regions. We are currently exploring new alternatives to identify the regions to weight and to learn their corresponding weights. Further experiments are also performed to evaluate our approach in the context of other OCT based applications such as fluid segmentation and using OCT scans from other vendors.

Acknowledgements. This work is funded by WWTF AugUniWien/FA7464A0249 (MedUniWien); VRG12-009 (UniWien). We thank NVIDIA Corporation for donating a GPU.

References

1. Breger, A., et al.: On orthogonal projections for dimension reduction and applications in variational loss functions for learning problems. arXiv:1901.07598 (2018)
2. Gerendas, B.S., et al.: OCT biomarkers predictive for visual acuity in patients with diabetic macular edema. IOVS **58**(8), 2026–2026 (2017)
3. Khened, M., Kollerathu, V.A., Krishnamurthi, G.: Fully convolutional multi-scale residual DenseNets for cardiac segmentation and automated cardiac diagnosis using ensemble of classifiers. Med. Image Anal. **51**, 21–45 (2019)
4. Litjens, G., et al.: A survey on deep learning in medical image analysis. Med. Image Anal. **42**, 60–88 (2017)
5. Orlando, J.I., et al.: U2-Net: a Bayesian U-Net model with epistemic uncertainty feedback for photoreceptor layer segmentation in pathological OCT scans. In: ISBI (2019)
6. Ronneberger, O., Fischer, P., Brox, T.: U-Net: convolutional networks for biomedical image segmentation. In: Navab, N., Hornegger, J., Wells, W.M., Frangi, A.F. (eds.) MICCAI 2015. LNCS, vol. 9351, pp. 234–241. Springer, Cham (2015). https://doi.org/10.1007/978-3-319-24574-4_28

7. Schmidt-Erfurth, U., Sadeghipour, A., Gerendas, B.S., Waldstein, S.M., Bogunović, H.: Artificial intelligence in retina. Prog. Retinal Eye Res. (2018). https://doi.org/10.1016/j.preteyeres.2018.07.004
8. Takahashi, A., et al.: Photoreceptor damage and reduction of retinal sensitivity surrounding geographic atrophy in age-related macular degeneration. Am. J. Ophthalmol. **168**, 260–268 (2016)

Foveal Avascular Zone Segmentation in Clinical Routine Fluorescein Angiographies Using Multitask Learning

Dominik Hofer[(✉)], José Ignacio Orlando, Philipp Seeböck, Georgios Mylonas,
Felix Goldbach, Amir Sadeghipour, Bianca S. Gerendas,
and Ursula Schmidt-Erfurth

Department of Ophthalmology and Optometry, Medical University of Vienna,
Vienna, Austria
dominik.hofer@meduniwien.ac.at

Abstract. Fluorescein Angiography (FA) is an imaging technique that allows to visualize the vascular structure of the retina. The Foveal Avascular Zone (FAZ) is a vessel-free area located at the center of the fovea whose shape characteristics are used to diagnose eye-related diseases such as diabetic retinopathy. Segmentation of the FAZ in FA therefore plays an important role in clinical decision making. However, manual delineation is costly and time-consuming. Current methods for automated FAZ segmentation either rely on segmenting the vasculature first, require manual initialization or are tailored to specific image properties. Hence, they often fail when dealing with images from clinical routine, which were usually acquired using multiple devices and at different imaging settings. In this paper we propose to overcome these limitations by means of a multitask learning approach. Our method exploits an additional Euclidean distance map prediction task to better deal with variable imaging conditions, by benefiting from its regularization effect. Our method is empirically evaluated using a data set of FA scans from large multicenter clinical trials with diverse qualities and image resolutions. The proposed model outperformed a baseline U-Net, achieving an average Dice of 0.805. To the best of our knowledge, our approach is the first deep learning method for FAZ segmentation in FA ever published.

1 Introduction

Fluorescein Angiography (FA) is a medical imaging technique that is widely applied to visualize the vascular structure of the retina [1]. Its acquisition process starts with the injection of a fluorescein dye to the blood stream in order to enhance the visibility of the vessels. While the dye reaches the relevant areas of the eye, a series of images (Fig. 3) are taken using a dye-sensitive camera. FA images are used for diagnosing diseases that are associated with the retinal vascular structure, such as diabetic retinopathy (DR) or age-related macular degeneration (AMD) [1].

© Springer Nature Switzerland AG 2019
H. Fu et al. (Eds.): OMIA 2019, LNCS 11855, pp. 35–42, 2019.
https://doi.org/10.1007/978-3-030-32956-3_5

Located within the center of the retina, the Foveal Avascular Zone (FAZ) is an area that is devoid of vessels and responsible for sharp, accurate vision [2]. Normally, the FAZ has a diameter of 0.4–0.5 mm [3]. Enlargements of the FAZ are associated with DR where the retina is damaged due to occlusion of perifoveolar capillaries or arterioles [2,3]. Hence, measuring the size, diameter and shape of the FAZ can aid clinicians to diagnose and monitor DR, which is usually done by manually analyzing FA images and segmenting the FAZ.

Technological advances in the area of medical image processing, computer vision and machine learning have led to the development of computer-aided systems that can automate the process of diagnosis and segmentation in ophthalmic imaging [4]. These systems can facilitate clinical decision making, reduce intra- and inter-observer variability, and improve accuracy, compared to manually performed tasks. Several methods have been introduced for FAZ segmentation, although most of them rely on manual initialization [5–8], explicit anatomical priors retrieved by other surrogate algorithms (e.g. blood vessel segmentation [6]) or standard image processing [9]. In contrast to these hand-crafted approaches, supervised deep learning techniques do not need manual initialization and are not limited to a-priori defined anatomical structures [10]. However, these methods were never applied before for FAZ segmentation.

Segmenting clinical routine images remains challenging even for conventional deep learning networks due to the large variation of image properties such as pixel resolution and size, field of views and contrast (Fig. 3). Furthermore, the size and shape of the FAZ can vary considerably depending on the age of the patient and the presence of pathologies (Fig. 3). In general, this difficulty can be alleviated by means of heavy data augmentation or by collecting abundant training data with different characteristics. However, selecting an appropriate combination of data augmentation strategies requires an intense engineering process to simulate every possible input. Moreover, manually annotating a large, diverse training set is also prohibitive. Alternatively, data can be standardized to a common representation to reduce the covariate shift between the training set and the real life data in which the model will be applied. However, this is also unfeasible when technical information regarding the acquisition process is missing.

In this paper we propose to overcome these issues by using a multitask learning approach. Our hypothesis is that representing the target information in the form of an additional prediction task can aid the learning process as a form of regularization, allowing the model to better characterize the inputs and therefore improve results. To this end, we adapted the model described in [11] and propose to use the prediction of the Euclidean distance map of the FAZ labels as an auxiliary task. We empirically validated our approach on a set of 96 test images from large multicenter clinical trials with diverse qualities and image resolutions. We observed that this method results in a model that has significantly less outliers and produces more compact, continuous segmentations of the FAZ. This approach might be useful when other information that is necessary for image standardization is not available (e.g., pixel size) or other generalization methods are not feasible.

2   Methods

2.1   Image Segmentation with Supervised Deep Learning

Given an input image $\mathbf{x}$, the goal is to predict a binary labeling $\hat{y}$ of the FAZ. In supervised learning, a training set $S = \{(x^{(i)}, y^{(i)}), 1 < i < n\}$ with n pairs of images $x^{(i)}$ and their corresponding manual annotations $y^{(i)}$ is used to train a deep learning model $F_\omega(x) = \hat{y}$ with parameters ω. This is achieved by minimizing a loss function $L(y, \hat{y})$ such as cross-entropy or mean squared error (MSE).

Depending on the characteristics of the training set, the segmentation models will converge to a specific solution that might not be general enough to be applied to any other input data. Furthermore, in challenging scenarios such as when dealing with clinical routine data, S is composed of images obtained with different devices, under variable settings and from patients with different pathologies. In this scenario, a-priori regularization techniques such as data augmentation are usually applied to improve generalization. However, the augmentation techniques must be carefully designed to avoid training using unrealistic images, which might affect the performance of the method when deployed. Additionally, modeling every possible image appearance is time-demanding and sometimes unfeasible.

2.2   FAZ Segmentation with Multitask Learning

In this paper we propose to overcome the difficulties of supervised deep learning models, when dealing with data from clinical routine, by regularization through multitask learning. We build on top of the segmentation framework proposed by Tan *et al.* [11] and adapt it to our specific domain. Contrary to a conventional U-Net, which has only a single encoder and decoder, such a multitask learning approach incorporates an additional decoder with its own set of parameters, which are learned separately for each branch, while sharing a single encoder between tasks.

Formally, in a multitask learning scenario a model $F_\omega(x) = \{\hat{y}, \hat{d}\}$ is trained to predict the FAZ labeling and its associated distance map for a given FA scan. The training data then becomes $S = \{(x^{(i)}, y^{(i)}, d^{(i)}), 1 < i < n\}$, where $d^{(i)}$ is the distance map associated to the manual labeling $y^{(i)}$. The distance map is computed by assigning each pixel the Euclidean distance to the nearest pixel of the FAZ labeling. Given this additional task, the model is trained using a weighted sum of losses from the two branches with specific parameters for each branch. More precisely, the model can be defined as $F_\omega(x) = \{\hat{y}, \hat{d}\} = \{f_{\omega_s, \omega_f}(x), g_{\omega_s, \omega_g}(x)\}$ with parameters $w = \{w_s, w_f, w_g\}$. w_s, w_f, w_g are the weights of the (shared) encoder, the mask decoder and the distance map decoder, and $f_{\omega_s, \omega_f}(x)$ and $g_{\omega_s, \omega_g}(x)$ are the two prediction tasks for the mask and distance map, respectively. The objective function is then a combination of two losses, defining a new risk minimization problem:

$$\hat{\omega} = \underset{\omega}{\mathrm{argmin}} \left\{ L_{cls}(y, f_{\omega_s, \omega_f}(x)) + \lambda * L_{dis}(d, g_{\omega_s, \omega_g}(x)) \right\} \tag{1}$$

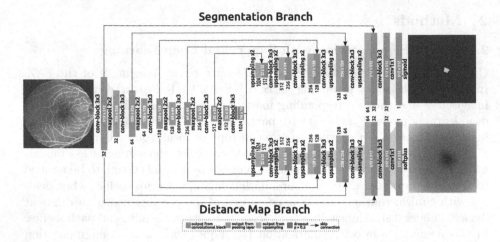

Fig. 1. Schematic representation of our method.

where $L_{cls}(y, \hat{y})$ is the cross-entropy between y and $\hat{y}$, $L_{dis}(d, \hat{d}) = ||d - \hat{d}||_2^2$ with $\hat{d}$ the predicted distance map and λ is a weight that balances the two losses.

In [11], a custom weighting function is also added to the loss function to further penalize errors that are further away from the segmentation border. However, this introduces a redundancy in the loss, as this has the same effect as estimating the differences between the distance maps. In our experiments we ignored such a weighting. Furthermore, we adapted the model in [11] to be based on a conventional U-Net [12] with batch normalization after each convolutional layer, dropout in the bottleneck layer (Fig. 1) and two decoders, one for each prediction task (the FAZ segmentation mask and the distance map). At test time, only the segmentation branch is used to retrieve a FAZ labeling.

3   Experimental Setup

3.1   Materials

To validate the performance of our approach to deal with heterogeneous data, we used a dataset comprising 494 FA images with sizes ranging from 4288×2848 to 512×512 pixels. The scans were taken with devices from seven different vendors. Figure 3 illustrates some exemplary images from the dataset. Notice that the field of view differs significantly in some of the images, which affects the variability of the pixel size (in μm), while others present croppings within specific anatomical regions and some are full images as obtained from the acquisition device. The manual labelings of the FAZ were provided by an experienced reader and used as ground truth. The dataset was randomly split into 318 images for the training set, 80 images for the validation set and 96 images for the test set. The test set was annotated independently by three other readers to assess the inter-reader variability by comparing the ground truth to each of the three additional readers.

As a preprocessing, all the images were cropped to eliminate borders as much as possible and resized to a resolution of 512 × 512 pixels. Due to the limited size of the training set, we performed standard data augmentation in the form of random horizontal flipping, rotation to a maximum of 10°, translation to a maximum of 10% of the image size, and zooming up to 156.25%. As a last step, a black circle was added around the borders of the image to unify the amount of information provided in each FA scan.

3.2 Evaluation Metrics

The performance was evaluated in the test set by calculating standard metrics such as the pixel-wise area under the Precision-Recall-Curve (AUC), Dice, precision and recall.

3.3 Training Details and Baselines

We trained our network for 500 epochs using Adam optimization and an initial learning rate of 10^{-4}. The model yielding the highest Dice on the validation set was then taken for final evaluation. Different values of $\lambda = 10^i, i \in \{-2, -1, 0, 1, 2\}$ were explored and evaluated on the validation set, with the model trained with $\lambda = 10$ being the best performing one. Finally, we trained a conventional U-Net as a baseline for comparison purposes in order to show that incorporating an additional task is useful. The baseline has the same configuration and is therefore equivalent to our network but with a single branch.

4   Results

Table 1 summarizes the quantitative results obtained by our multitask based approach and the baseline, in terms of precision, recall and Dice. The multitask model is observed to outperform the baseline in terms of precision and Dice. The U-Net reported the highest sensitivity but at a significantly lower precision value, which indicates the presence of several false positives. This behavior is also observed in Fig. 2 (left), which depicts pixel-wise precision/recall curves obtained from the score maps produced by each approach. The multitask model reported higher precision values than the baseline when analyzed at the same recall values, resulting in an AUC of 0.854 compared to 0.835 reported by the standard U-Net.

Figure 2 (right) includes a boxplot showing the distribution of Dice values obtained by each method and the three independent readers. Our multitask approach achieved a median Dice of 0.870 compared to 0.852 obtained by the standard U-Net. A decrease in the number of low-performing images is also observed in the figure, with the multitask model reporting consistently less outliers.

Figure 3 qualitatively presents exemplary segmentations obtained by our approach, the baseline U-Net and the corresponding manual annotations. It can be

Table 1. Mean ± standard deviation of recall, precision and Dice for the baseline U-Net, our approach, and the three manual reader annotations.

Model	Precision	Recall	Dice
U-Net	0.753 ± 0.202	**0.871 ± 0.203**	0.782 ± 0.179
Multitask Net	**0.826 ± 0.204**	0.815 ± 0.199	**0.805 ± 0.180**
Reader 1	0.923 ± 0.081	0.846 ± 0.131	0.874 ± 0.097
Reader 2	0.943 ± 0.071	0.836 ± 0.127	0.878 ± 0.092
Reader 3	0.883 ± 0.111	0.868 ± 0.099	0.867 ± 0.075

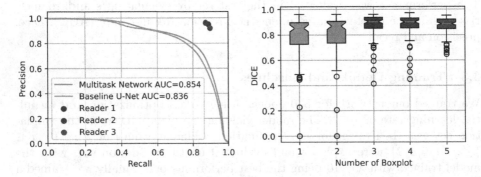

Fig. 2. Left: Precision-Recall-Curves for the segmentations. Right: Boxplots of the Dice values. (1) Baseline, (2) Multitask Network, (3–5) manual reader annotations.

observed that the multitask model improves the appearance of the final segmentations on a diverse set of heterogeneous images. In the top-left image, our approach generates a more continuous, coherent shape for the FAZ than the U-Net. The top-right image shows an angiography where the vessels cannot be clearly seen (contrary to the other three images in the figure). Due to the low contrast, the baseline U-Net tends to oversegment the FAZ area. In this case, the multitask approach aids to reduce the size of the segmentation and better fit the shape of the actual FAZ. The bottom-left FA scan is an example of how small outliers are deleted when using the auxiliary branch. Finally, the bottom-right image illustrates that our approach can also manage blurred, low-quality image with a pathologically increased FAZ.

5 Discussion

In this paper we propose to apply multitask deep supervised learning for FAZ segmentation in diverse FA images from large multicenter clinical trials. When working with such heterogeneous scans, data standardization is crucial to reduce the covariate shift between training and test images. Since this is usually unfeasible when dealing with clinical routine data (e.g. due to unknown imaging settings), the alternative is to impose some form of regularization to the model

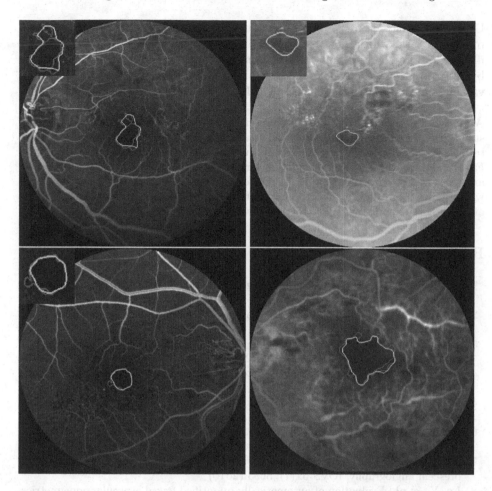

Fig. 3. Qualitative comparison of the segmentation result in a set of heterogeneous FA scans with different contrast, sharpness, field-of-view, and brightness. Red: Manual reader annotation (ground truth). Blue: Baseline U-Net. Cyan: Our multitask approach. (Color figure online)

to improve its generalization ability. Heavy data augmentation, for instance, is the most straightforward technique to learn generic models. However, this has limitations as it requires to manually craft and model every possible imaging setting. Instead, we propose to improve generalization by incorporating a multitask method with an auxiliary second branch that predicts the Euclidean distance map computed from the FAZ labeling. By means of this additional decoder, our model shows to better generalize in terms of appearance and localization of the region of interest. Incorporating this second task is beneficial on heterogeneous data for which information such as the pixel size is missing. We experimentally observed that our multitask approach is better than a standard

U-Net in this scenario. We hypothesize this is due to the inductive bias of the regression task, which aids the segmentation branch and acts as a form of regularization. Further research is being done to improve the segmentation results by incorporating an adversarial loss and examining the performance in other vascular imaging modalities such as OCT-A.

Acknowledgements. The authors would like to express their sincere gratitude to all staff of the Vienna Reading Center, who contributed to the image ground truth annotation. This study was in part funded by the Austrian Federal Ministry of Economy, Family and Youth, National Foundation for Research, Technology and Development and by Novartis (Christian Doppler Laboratory for Ophthalmic Image Analysis). Data was acquired from the Vienna Reading Center Image Database and as such we thank all investigators and study sponsors who have participated to form this database.

References

1. Abràmoff, M.D., et al.: Retinal imaging and image analysis. IEEE Rev. Biomed. Eng. **3**, 169–208 (2010)
2. Conrath, J., et al.: Foveal avascular zone in diabetic retinopathy: quantitative vs qualitative assessment. Eye **19**(3), 322 (2005)
3. Bresnick, G., et al.: Abnormalities of the foveal avascular zone in diabetic retinopathy. Arch. Ophthalmol. **102**(9), 1286–1293 (1984)
4. Schmidt-Erfurth, U., et al.: Artificial intelligence in retina. Prog. Retinal Eye Res. **67**, 1–29 (2018)
5. Haddouche, A., et al.: Detection of the foveal avascular zone on retinal angiograms using Markov random fields. Digit. Signal Process. **20**(1), 149–154 (2010)
6. Fadzil, M.A., et al.: Determination of foveal avascular zone in diabetic retinopathy digital fundus images. Comput. Biol. Med. **40**(7), 657–664 (2010)
7. Zheng, Y., et al.: Automated segmentation of foveal avascular zone in fundus fluorescein angiography. IOVS **51**(7), 3653 (2010)
8. Lu, Y., et al.: Evaluation of automatically quantified foveal avascular zone metrics for diagnosis of diabetic retinopathy using optical coherence tomography angiography. IOVS **59**(6), 2212 (2018)
9. Díaz, M., et al.: Automatic segmentation of the foveal avascular zone in ophthalmological OCT-A images. PloS One **14**(2), e0212364 (2019)
10. Litjens, G., et al.: A survey on deep learning in medical image analysis. MedIA **42**, 60–88 (2017)
11. Tan, C., et al.: Deep multi-task and task-specific feature learning network for robust shape preserved organ segmentation. In: ISBI 2018, pp. 1221–1224 (2018)
12. Ronneberger, O., Fischer, P., Brox, T.: U-Net: convolutional networks for biomedical image segmentation. In: Navab, N., Hornegger, J., Wells, W.M., Frangi, A.F. (eds.) MICCAI 2015. LNCS, vol. 9351, pp. 234–241. Springer, Cham (2015). https://doi.org/10.1007/978-3-319-24574-4_28

Guided M-Net for High-Resolution Biomedical Image Segmentation with Weak Boundaries

Shihao Zhang[1], Yuguang Yan[1], Pengshuai Yin[1], Zhen Qiu[1], Wei Zhao[2], Guiping Cao[2], Wan Chen[3], Jin Yuan[3], Risa Higashita[4], Qingyao Wu[1], Mingkui Tan[1(✉)], and Jiang Liu[5,6]

[1] South China University of Technology, Guangzhou, China
[2] CVTE Research, Guangzhou, China
[3] Zhongshan Ophthalmic Center Sun Yat-sen University, Guangzhou, China
[4] Tomey Corporation, Nagoya, Japan
[5] Southern University of Science and Technology, Shenzhen, China
[6] Cixi Institute of BioMedical Engineering, Ningbo Institute of Industrial Technology, Chinese Academy of Sciences, Beijing, China

Abstract. Biomedical image segmentation plays an important role in automatic disease diagnosis. However, some particular biomedical images have blurred object boundaries, and may contain noises due to the limited performance of imaging device. This issue will highly affects segmentation performance, and will become even severer when images have to be resized to lower resolution on a machine with limited memory. To address this, we propose a guide-based model, called G-MNet, which seeks to exploit edge information from guided map to guide the corresponding lower resolution outputs. The guided map is generated from multi-scale input to provide a better guidance. In these ways, the segmentation model will be more robust to noises and blurred object boundaries. Extensive experiments on two biomedical image datasets demonstrate the effectiveness of the proposed method.

1 Introduction

Biomedical image segmentation plays important role in automatic disease diagnosis. In particular, in glaucoma screening, correct optic disc (OD) and optic cup (OC) segmentation will help obtain an accurate vertical cup-to-disc ratio (CDR), which is commonly used for glaucoma diagnosis. Moreover, in cataract grading, lens structure segmentation helps to calculate the density of different lens parts, and the density quantification is a kind of cataract grading metric [11].

In recent years, Convolutional neural networks (CNNs) have shown strong power in biomedical image segmentation with remarkable accuracy. For example, [9] proposes a U-shape convolutional network (U-Net) to segment images

---

This work was done when S. Zhang and Y. Yan are interns at CVTE Research.

H. Fu et al. (Eds.): OMIA 2019, LNCS 11855, pp. 43–51, 2019.
https://doi.org/10.1007/978-3-030-32956-3_6

with precise boundaries by constructing skip connections to restore the information loss caused by pooling layers. [5] proposes an M-shape convolutional network, which combines multi-scale inputs and constructs local outputs to link the loss and early layers. In practice, however, some high-resolution biomedical images have noises and blurred boundaries, like the anterior segment optical coherence tomography (AS-OCT) images, which may hamper the segmentation performance, as shown in Fig. 1. Furthermore, suffering from the limitation of memory, existing methods usually receive down-sampled images as input and then up-sample the results back to the original resolution, which, however, may lead even worse segmented boundaries.

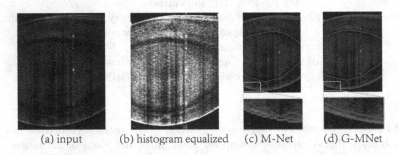

(a) input   (b) histogram equalized   (c) M-Net   (d) G-MNet

Fig. 1. (a): An AS-OCT image sample with weak nucleus and cortex boundaries. (b): corresponding histogram equalized image with a lot of noise. (c): segmentation results of M-Net with low-resolution input. (d): segmentation results of G-MNet.

To address the above issues and hence improve the segmentation performance, we seek to exploit guided filter to extract edge information from high-resolution images. In this way, high-quality segmentation results can be generated from low-resolution poorly segmented results. Moreover, precise segmented boundaries can be maintained after up-sampling. Guided filter [6] is an edge-preserving image filter and has been incorporated into deep learning on several tasks. For example, [12] formulates it into an end-to-end trainable module, [7] combines it with superpixels to decrease computational cost. Different from existing works which use guided filter as post-processing, we incorporate the guided filter into CNNs to learn better features for segmentation.

Unfortunately, the performance of the guided filter will be affected by noises and blurred boundaries in images. Therefore, better guidance rather than the original image is required. In this sense, we design a guided block to produce an informative guided map, which helps to alleviate the influence of noises and blurred boundaries. Besides, multi-scale features and multi-scale inputs are also combined to make model more robust to noise. Thorough experiments on two benchmark datasets, namely CASIA-2000 and ORIGA datasets, demonstrate the effectiveness of our method. Our method also achieves the best performance on CASIA-2000 dataset and outperforms the state-of-the-art OC and/or OD segmentation methods on ORIGA dataset.

2 Methodology

In this section, we provide an overview of our guide-based model, named G-MNet, in Fig. 2. Then introduce its three components: an M-shape convolutional network (M-Net) to learn hierarchical representations, a guided block for better guidance, and a multi-guided filtering layer to filter multi-scale low-resolution outputs. Our G-MNet firstly generates multi-scale side-outputs by M-Net, then these side-outputs are filtered to high-resolution through the multi-guided filtering layer. The guided block is exploited to provide better guidance for the multi-guided filtering layer. After that, an average layer is employed to combine all the high-resolution outputs. At last, the multi-guided filter receives the combined outputs and produces the final segmentation result.

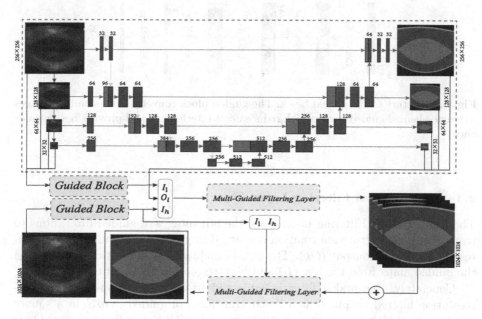

Fig. 2. Overview of the proposed deep architecture. Firstly, multi-scale side-outputs are generated by M-Net. Then the multi-guided filtering layer filters these side-output to high-resolution with the guidance from the guided map. At last, an average layer is employed to combine all the outputs, and the result is then guided to produce the final segmented output.

2.1 M-Shape Convolutional Network

We choose the M-Net [5] as the main body of our method, as shown by the red dashed box in Fig. 2. The M-Net includes a U-Net used to learn a rich hierarchical representation. Besides, multi-scale input and side-output are combined to better leverage multi-scale information.

2.2 Guided Block

In order to provide better guidance and reduce the impact of noise, we design a guided block to produce guided maps. The guided maps contain the main structure information extracted from the original images and also remove the noisy components. Figure 3 shows the architecture of the guided block. The guided block contains two convolution layers, between which are an adaptive normalization layer and a leaky ReLU layer. After the second convolution layer, an adaptive normalization layer [3] is added. The guided block is jointly trained with the entire network, thus the produced guided maps cooperate better with the rest of the model compared with the original image.

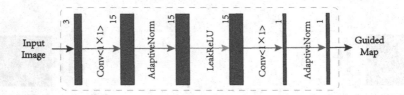

Fig. 3. Structure of the guided block. The guided block converts three-channel images to single-channel guided maps which reduce noise interference and provide better guidance.

2.3 Multi-guided Filtering Layer

The Multi-Guided Filtering Layer, take the advantages of guided filter, aims to transform the structure information contained in guided map and produce high-resolution filtered output (O_h). The inputs includes low-resolution output (O_l) the guided maps from the low (I_l) and high-resolution (I_h) input image.

Concretely, the guided filter is subjected to an assumption that the low-resolution filtered output $\hat{O}$ is a linear transform of guided map I in a square window w_k, which is centered at the position k with the radius being r. O_h is up-sampled from $\hat{O}$. The definition of $\hat{O}$ with respect to w_k is given as:

$$\hat{O}_{ki} = a_k I_{l_i} + b_k, \forall i \in w_k, \tag{1}$$

where (a_k, b_k) are some linear coefficients assumed to be constant in w_k and the radius of window is r.

a_k, b_k can be obtained by minizing the loss function:

$$E(a_k, b_k) = \sum_{i \in w_k} ((a_k I_{l_i} + b_k - O_{l_i})^2 + \epsilon a_k^2), \tag{2}$$

where ϵ is a regularization parameter penalizing large a_k.

Considering that each position i is involved in multiple windows $\{w_k\}$ with different coeffecients $\{a_k, b_k\}$, we average all the values of $\hat{O}_{ki}$ from different windows to generate $\hat{O}_i$, which is equal to average the coefficients (a_k, b_k) of all the windows overlapping i, i.e.,

$$\hat{O}_i = \frac{1}{N_k} \sum_{k \in \Omega_i} a_k I_{l_i} + \frac{1}{N_k} \sum_{k \in \Omega_i} b_k = A_{l_i} * I_{l_i} + B_{l_i}, \tag{3}$$

where Ω_i is the set of all the windows including the position i, and $*$ is the element-wise multiplication. After upsampling A_l and B_l to obtain A_h and B_h, respectively, the final output is calcuted as (Fig. 4):

$$O_h = A_h * I_h + B_h. \tag{4}$$

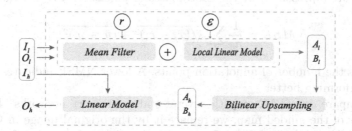

Fig. 4. Illustrations of multi-guided filtering layer. With low-resolution input I_l, O_l and hyperparameters r, ϵ, low-resolution A_l, B_l are calculated. By bilinear upsampling A_l, B_l, high-resolution A_h, B_h are generated which are then used to produce the final high-resolution output O_h with high-resolution guided map I_h.

3 Experiments

3.1 Datasets

(1) CASIA-2000: We collect high-resolution AS-OCT images with weak boundaries and noise from CASIA-2000 produced by Tomey Co. Ltd. The dataset contains 2298 images, including 1711 training images and 587 testing images. All the images are annotated by experienced ophthalmologists.

(2) ORIGA: It contains 650 fundus images with 168 glaucomatous eyes and 482 normal eyes. The 650 images are divided into 325 training images (including 73 glaucoma cases) and 325 testing images (including 95 glaucomas).

3.2 Training Details

We train our G-MNet from scratch for 80 epochs using Adam optimiser with the learning rate being 0.001. For the experiments on CASIA-2000 dataset, we set $\epsilon = 0.01$ and $r = 5$. The original image size is 2130×1864. We crop the lens area, which is about 1024×1024 pixels, and resize it into 1024×1024 and 256×256 for high- and low-resolution inputs. For the experiments on ORIGA dataset, we set $\epsilon = 0.9$ and $r = 2$. The original image size is 3072×2048. We train a LinkNet [1] on training set to crop the OD area, and then resize it into 256×256 for low-resolution inputs.

3.3   Results on CASIA-2000 Dataset

Segmentation on CASIA-2000 aims to evaluate capsule, cortex and nucleus segmentation performance. Following the previous work in AS-OCT image segmentation [15], we employ the normalized mean squared error (NMSE) between a predicted shape $S_p = \{\hat{x}_i, \hat{y}_i\}$ and the ground truth shape $S_g = \{x_i, y_i\}$, where the shapes are represented by the coordinates of pixels. NMSE is defined as

$$NMSE = \frac{1}{n_g} \sum_{i=1}^{n_g} \sqrt{(\hat{x}_i - x_i)^2 + (\hat{y}_i - y_i)^2}, \qquad (5)$$

where n_g is the number of annotation points. A lower NMSE indicates the network is performing better.

We compare our G-MNet with several state-of-the-art networks. To verify the efficacy of the guided map, we replace it by the original image in G-MNet, and named this model **G-MNet-Image**. To test the performance of guiding in multi-scale, we construct a special G-MNet, named **G-MNet-Single**, which only filters the final averaged result without filtering multi-scale side-outputs. Table 1 shows the performance of different methods. We have the following observations: Firstly, G-MNet-Single performs better than M-Net, which indicates that guided filter is able to improve the accuracy of segmentation. Secondly, G-MNet outperforms G-MNet-Single by 0.16, 0.20 and 0.17 in capsule, cortex and nucleus boudary, respectively. This demonstrates the effectiveness of the learning strategy in multi-scale. Lastly, G-MNet performs much better than G-MNet-Image, which is disturbed by noises. This verifies that guided maps are able to provide better guidance for reducing noises.

Table 1. Segmentation results on CASIA-2000.

Method	Capsule	Cortex	Nucleus
FCN-VGG16 [8]	3.08 ± 4.84	3.34 ± 3.14	11.03 ± 4.08
DeepLabV2-Res101 [2]	3.97 ± 4.08	6.18 ± 4.31	10.88 ± 8.04
PSPNet-Res34 [16]	1.37 ± 0.96	1.73 ± 0.75	8.20 ± 3.97
M-Net [5]	1.37 ± 2.62	1.60 ± 0.93	7.93 ± 3.65
G-MNet-Image (ours)	3.23 ± 1.46	4.39 ± 1.34	9.44 ± 2.75
G-MNet-Single (ours)	0.73 ± 0.72	1.17 ± 0.91	7.62 ± 3.29
G-MNet (ours)	$\mathbf{0.57 \pm 0.29}$	$\mathbf{0.97 \pm 0.60}$	$\mathbf{7.45 \pm 3.24}$

3.4   Results on ORIGA Dataset

Following the previous work [5], we evaluate the OD and/or OC segmentation performance and employ the following overlapping error (OE) as the evaluation metric:

$$OE = 1 - \frac{A_{GT} \bigcap A_{SR}}{A_{GT} \bigcup A_{SR}}, \tag{6}$$

where A_{GT} and A_{SR} denote the areas of the ground truth and segmented mask, respectively.

We compare our G-MNet to the state-of-the-art methods in OD and/or OC segmentation, including ASM [14], SP [4], SW [13], U-Net [9], M-Net [5], M-Net with polar transformation (M-Net + PT) and Sun's [10].

Following the setting in [5], we firstly localize the disc center, and then crop 640×640 pixels to obtain the input images. Inspired by M-Net+PT, Inspired by M-Net+PT [5], we provide the results of G-MNet with polar transformation, called G-MNet+PT. Besides, to reduce the impacts of changes in the size of OD, we construct a method G-MNet+PT+50, which enlarges 50 pixels of bounding-boxes in up, down, right and left, where the bounding boxes are obtained from our pretrained LinkNet.

Table 2. Segmentation results on ORIGA.

Method	OE_{disc}	OE_{cup}
ASM [14]	0.148	0.313
SP [4]	0.102	0.264
SW [13]	–	0.284
Sun's [10]	0.069	0.213
U-Net [9]	0.115	0.287
M-Net [5]	0.083	0.256
M-Net+PT [5]	0.071	0.230
G-MNet (ours)	0.075	0.229
G-MNet+PT (ours)	0.069	0.213
G-MNet+PT+50 (ours)	**0.062**	**0.211**

Table 2 shows the segmentation results, the overlapping errors of other approaches come directly from the published results. Our method outperforms all the state-of-the-art OD and/or OC segmentation algorithms in terms of the aforementioned two evaluation criteria, which demonstrates the effectiveness of our model. Besides, Our G-Mnet outperforms M-Net by 0.008 and 0.027 in OE_{disc} and OE_{cup}, respectively. Simultaneously, Our G-Mnet+PT also performs better than M-Net+PT. These results indicate that our modification to M-Net has a great help to the performance.

4   Conclusions

In this paper, we propose a guide-based M-shape convolutional network, G-MNet, to segment biomedical images with weak boundaries, noise and high-resolution. Our G-MNet products high-quality segmentation results by incorporating guided filter into CNNs to learn better features for segmentation. It also benefit from the informative guided maps which provide better guidance and reduce the influence of noise by extracting the main feature from the original images. We further filter multi-scale side-outputs to construct the guided block more robust to noise and scaling. Thorough epxeriment on two benchmark datasets demonstrate the effectiveness of our method.

Acknowledgements. This work was supported by National Natural Science Foundation of China (NSFC) 61602185 and 61876208, Guangdong Introducing Innovative and Enterpreneurial Teams 2017ZT07X183, and Guangdong Provincial Scientific and Technological Fund 2018B010107001, 2017B090901008 and 2018B010108002, and Pearl River S&T Nova Program of Guangzhou 201806010081, and CCF-Tencent Open Research Fund RAGR20190103, and National Key R&D Program of China #2017YFC0112404.

References

1. Chaurasia, A., Culurciello, E.: LinkNet: exploiting encoder representations for efficient semantic segmentation. In: VCIP. IEEE (2017)
2. Chen, L.C., et al.: DeepLab: semantic image segmentation with deep convolutional nets, atrous convolution, and fully connected CRFs. TPAMI **40**, 834–848 (2018)
3. Chen, Q., et al.: Fast image processing with fully-convolutional networks. In: ICCV (2017)
4. Cheng, J., et al.: Superpixel classification based optic disc and optic cup segmentation for glaucoma screening. TMI **32**, 1019–1032 (2013)
5. Fu, H., et al.: Joint optic disc and cup segmentation based on multi-label deep network and polar transformation. TMI **37**, 1597–1605 (2018)
6. He, K., et al.: Guided image filtering. TPAMI **35**, 1397–1409 (2013)
7. Hu, P., et al.: Deep level sets for salient object detection. In: CVPR (2017)
8. Long, J., et al.: Fully convolutional networks for semantic segmentation. In: CVPR (2015)
9. Ronneberger, O., Fischer, P., Brox, T.: U-Net: convolutional networks for biomedical image segmentation. In: Navab, N., Hornegger, J., Wells, W.M., Frangi, A.F. (eds.) MICCAI 2015. LNCS, vol. 9351, pp. 234–241. Springer, Cham (2015). https://doi.org/10.1007/978-3-319-24574-4_28
10. Sun, X., et al.: Localizing optic disc and cup for glaucoma screening via deep object detection networks. In: Stoyanov, D., et al. (eds.) OMIA/COMPAY -2018. LNCS, vol. 11039, pp. 236–244. Springer, Cham (2018). https://doi.org/10.1007/978-3-030-00949-6_28
11. Wong, A.L., et al.: Quantitative assessment of lens opacities with anterior segment optical coherence tomography. Br. J. Ophthalmol. **93**, 61–65 (2009)
12. Wu, H., et al.: Fast end-to-end trainable guided filter. In: CVPR (2018)

13. Xu, Y., et al.: Sliding window and regression based cup detection in digital fundus images for glaucoma diagnosis. In: Fichtinger, G., Martel, A., Peters, T. (eds.) MICCAI 2011. LNCS, vol. 6893, pp. 1–8. Springer, Heidelberg (2011). https://doi.org/10.1007/978-3-642-23626-6_1
14. Yin, F., et al.: Model-based optic nerve head segmentation on retinal fundus images. In: EMBC. IEEE (2011)
15. Yin, P., et al.: Automatic segmentation of cortex and nucleus in anterior segment OCT images. In: Stoyanov, D., et al. (eds.) OMIA/COMPAY -2018. LNCS, vol. 11039, pp. 269–276. Springer, Cham (2018). https://doi.org/10.1007/978-3-030-00949-6_32
16. Zhao, H., et al.: Pyramid scene parsing network. In: CVPR, pp. 2881–2890 (2017)

3D-CNN for Glaucoma Detection Using Optical Coherence Tomography

Yasmeen George[1]([✉]), Bhavna Antony[1], Hiroshi Ishikawa[2], Gadi Wollstein[2], Joel Schuman[2], and Rahil Garnavi[1]

[1] IBM Research - Australia, Melbourne, VIC, Australia
georgey@ibm.com
[2] NYU Langone Eye Center, NYU School of Medicine, New York City, USA

Abstract. The large size of raw 3D optical coherence tomography (OCT) volumes poses challenges for deep learning methods as it cannot be accommodated on a single GPU in its original resolution. The direct analysis of these volumes however, provides advantages such as circumventing the need for the segmentation of retinal structures. Previously, a deep learning (DL) approach was proposed for the detection of glaucoma directly from 3D OCT volumes, where the volumes were significantly downsampled first. In this paper, we propose an end-to-end DL model for the detection of glaucoma that doubles the number of input voxels of the previously proposed method, and also boasts an improved AUC = 0.973 over the results obtained using the previously proposed approach of AUC = 0.946. Furthermore, this paper also includes a quantitative analysis of the regions of the volume highlighted by grad-CAM visualization. Occlusion of these highlighted regions resulted in a drop in performance by 40%, indicating that the regions highlighted by gradient-weighted class activation maps (grad-CAM) are indeed crucial to the performance of the model.

Keywords: 3D-CNN · Glaucoma detection · Optical coherence tomography · Gradient-weighted class activation maps · Visual explanations

1 Introduction

Glaucoma is the second leading cause of irreversible blindness worldwide. The number of worldwide glaucoma patients, aged 40–80 years, is estimated to be approximately 80 million in 2020 with about 20 million increase since 2010 [4]. This disease is characterised by optic nerve damage, the death of retinal ganglion cells [3], and the ultimate loss of vision. It is a slowly progressing disease, with a long asymptomatic phase, where patients do not notice the increasing loss of peripheral vision. Since glaucomatous damages are irreversible, early detection is crucial.

H. Fu et al. (Eds.): OMIA 2019, LNCS 11855, pp. 52–59, 2019.
https://doi.org/10.1007/978-3-030-32956-3_7

Spectral-domain OCT imaging provides clinicians with high-resolution images of the retinal structures, which are employed for diagnosing and monitoring retinal diseases, evaluating progression, and assessing response to therapy [5]. While the previous approaches around the detection of glaucoma have primarily depended on segmented features such as the thickness of the retinal nerve fibre layer (RNFL) and the ganglion cell layer (GCL), there have been limited efforts around evaluating the utility of deep learning (DL) models in improving the diagnostic accuracy and early detection of glaucoma using 3D OCT scans.

For example, in [2] and [16], DL networks were proposed to diagnose early glaucoma using retinal thickness features. Similarly in [11] and [1], pretrained models (trained on ImageNet [14]) were used for the detection task. In fact, none of these techniques use the raw volumes for DL training. Rather, they rely entirely on segmented features or measurements generated by the SD-OCT scanners. One limitation for this is the segmentation error propagation where failure rate increases with the disease severity and co-existing pathologies. This also does not allow the diagnosis model to learn other unknown features existing within the image data. The only end-to-end DL model which uses the 3D raw scans was proposed by Maetschke et al. [9] (referred to as CAM-3D-CNN). This approach utilised 3D convolutional layers in the CNN, but was forced to downsample the volumes by nearly a factor of 80 to enable CAM and train the model on a GPU (due to memory constraints on the GPU itself).

Another important aspect of DL is the clinical interpretability and transparency [12] of the models developed. Class activation mapping (CAM) [18] and gradient-weighted class activation maps (grad-CAM) [13] have been recently proposed to reveal insights into the decisions of deep learning models. Both of these techniques identify areas of the images that the networks relied on heavily to generate the classification. However, CAMs requires a specific network architecture, namely the use of a global average pooling layer prior to the output layer. Grad-CAM is a generalized form of CAM and can be used with any CNN-based architecture without any additional requirements. In this regard, the visualization of DL model for glaucoma detection has been studied in two papers [1,9]. An et al. [1] identified pathologic regions in 2D thickness maps using grad-CAM [13], which have shown to be in agreement with the important decision making regions used by physicians. Similarly, Maetschke et al. [9] implemented 3D-CAM [18] to identify the important regions in 3D OCT volumes. The maps were however, in a coarse resolution that matched the downsampled input image. This method also employed specific architecture changes to accommodate the requirements of CAM generation. It is also noteworthy that neither of these approaches analysed the CAMs in any systematic fashion, and merely used the heat maps to validate findings in a small number of images that were qualitatively assessed.

In this paper, we propose an end-to-end 3D-CNN for glaucoma detection trained directly on 3D OCT volumes (gradCAM-3D-CNN). This approach continues to avoid the dependency on segmented structural thicknesses, but also improves on previously approached techniques by doubling the size of the input volumes [9], and also improves on the performance in a direct comparison between gradCAM-3D-CNN and CAM-3D-CNN models. The use of 3D grad-CAM [13] allows for the visualization of the important regions of the 3D OCT cubes in a higher resolution than was not available before. Crucially, we validate the grad-CAM heat maps in a quantitative fashion, by occluding regions identified in the heat maps and assessing the impact of this on the performance of the model.

2   Materials and Methods

2.1   Dataset

The dataset contained 1248 OCT scans from both eyes of 624 subjects, acquired on a Cirrus SD-OCT Scanner (Zeiss; Dublin, CA, USA). 138 scans with signal strength less than 7 were discarded. The final dataset contained 263 scans on healthy eyes and 847 scans with primary open angle glaucoma (POAG). The scans were centered on the optic nerve heard (ONH) and had $200 \times 200 \times 1024$ (a-scans $\times$ b-scans $\times$ depth) voxels per cube covering an area of $6 \times 6 \times 2$ mm^3.

2.2   Network Architecture

The proposed CNN model receives input scans with a resolution of $256 \times 128 \times 128$ (depth $\times$ b-scans $\times$ a-scans) to classify an OCT volume as healthy or glaucoma. The network consists of eight 3D-convolutional layers, where each is followed by ReLU activation [6], batch-normalization [8] and max-pooling in order. The 3D convolutional layers have incremental number of the filters of 16-16-32-32-32-32-32-32 with kernel sizes of 3-3-3-3-5-5-3-3 in order, and stride of 1 for all layers. Also, 3D max-pooling layers has size of 2 and stride of 2. Finally, two fully-connected layers connect all the activated neurons in the previous layer to the next layer with 64 and 2 units respectively.

2.3   Evaluation: Training and Testing

The 1110 OCT volumes were downsampled to size $256 \times 128 \times 128$ and split into a training, validation and testing subsets, containing 889 (healthy: 219, POAG: 670), 111 (healthy: 23, POAG: 88) and 110 (healthy: 21, POAG: 89) scans, respectively. The proposed 3D-CNN model was trained using the RMSprop optimizer with a learning rate of $1e^{-4}$. Training was performed with a batch of size four through 50 epochs. Data was stratified per epoch by down-sampling to obtain balanced training samples. After each epoch, the area under the curve (AUC) was computed for the validation set and the network is saved if an improvement in AUC is observed.

2.4 DL Visualization

We implemented 3D grad-CAM [13] for visual explanations of the proposed model. We do not use CAM as it requires adding the global average pooling (GAP) after the last convolutional layer (i.e. conv#8) which restricts the network architecture design. Further, CAM would generate visualization only for feature map of conv#8, which in our case has a size of $2 \times 1 \times 1$. Hence, when resizing and overlaying on the original cube of size $200 \times 200 \times 1024$ will not provide any meaningful results. In this paper, we calculated the heat map for each of the first 6 convolutional layers (conv#1-6) separately, following the explanation provided in [13]. We did not compute grad-CAM for conv #7 and #8 layers due to the very small size of the corresponding heat maps (conv#7: $8 \times 4 \times 4$ and conv#8: $4 \times 2 \times 2$). The generated heat maps have the same size as the feature map of the corresponding convolutional layer. Instead of clipping the negative values in the resulted heat maps, as performed in the grad-CAM paper [13], we used the absolute value. To get rid of noisy gradients and to highlight only the important decision regions, we clipped the smallest 30% values and then resize the heat map to the original cube size.

To validate the generated heat maps, we occluded the input volumes by zeroing the rows and columns with the highest weights. Specifically, we extracted a set of indices with the highest weights per each dimension. This was done by spatial dimension reduction using average pooling. For example, a heat map with size $1024 \times 200 \times 200$ was reduced to a vector of size $1024 \times 1 \times 1$ by averaging the values of each 200×200 map to get a single value. The indices of the top highest values (top x) in the resulted vector represent the most important region for this dimension. We applied this process on the b-scans and depth dimensions with x values of 64 and 256 respectively, while we considered the 200 a-scan columns were all important. This means that a fixed region of size $256 \times 64 \times 200$ was occluded for each volume. Finally, the network was examined by evaluating the performance using the test set and its occluded volumes (2×110 scans).

3 Experiments and Results

3.1 The Glaucoma Detection Model

The proposed gradCAM-3D-CNN model as well as the CAM-3D-CNN model described in [9] were implemented using Python, Keras with Tensorflow [7] and nuts-flow/ml [10] on a single K80 GPU. Performance of both models were evaluated using five statistical measures, namely, area under the curve (AUC), accuracy, Matthews correlation coefficient (MCC), recall, precision and F1-score. We computed the weighted average measures to avoid biased resulting from the class size imbalance in the data. The threshold with the highest validation F1-score was chosen for calculating the performance measures. The proposed model achieved an AUC of 0.973 for the test set (110 scans).

Further, to validate the performance of the proposed model (gradCAM-3D-CNN), we trained the CAM-3D-CNN architecture, proposed in [9] using same

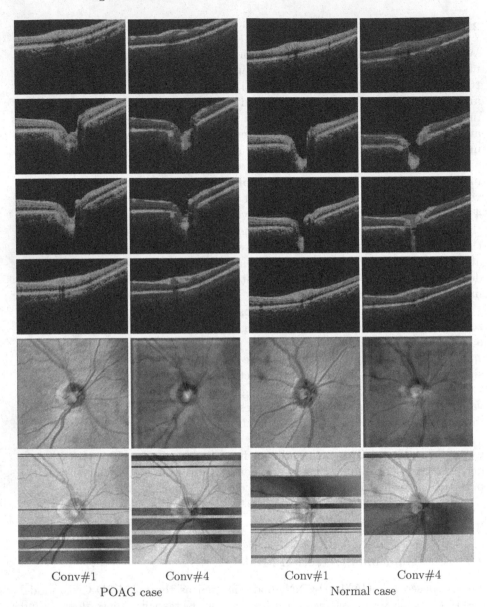

Conv#1   Conv#4   Conv#1   Conv#4
        POAG case   Normal case

Fig. 1. 3D grad-CAM visualization results. Rows 1–4 show the b-scan slices #50, #100, #110, and #140 in order; 5th row displays the enface of the overlay of grad-CAM heat map on the original 3D cube; 6th row displays the enface of the occluded region (refer to Sect. 2.4 for the occlusion method). Note: scans are resized for display.

data split. Table 1 has the performance measures for each model using the same test set. The table shows that the proposed model outperforms the CAM-3D-CNN model with an increase of 3%, 5%, and 9% in the AUC, accuracy and F1-

Table 1. Performance measures of the proposed model and the literature 3D-CNN model [9]

	Val. Thresh.	Accuracy	MCC	Recall	Precision	F1-score	AUC
Proposed network	0.394	0.923	0.879	0.964	0.963	0.963	0.973
3D-CNN network [9]	0.424	0.879	0.657	0.864	0.902	0.873	0.946

Table 2. Occlusion results for CAM and gradCAM heat maps (AUC of original model is 0.973)

	Map size	Accuracy	MCC	Recall	Precision	F1-score	AUC
GradCAM-Conv6-lyr19[a]	$16 \times 8 \times 8$	0.520	0.032	0.459	0.705	0.509	0.596
GradCAM-Conv4-lyr11[a]	$64 \times 32 \times 32$	0.534	0.053	0.495	0.713	0.546	0.624
GradCAM-Conv5-lyr15[a]	$32 \times 16 \times 16$	0.534	0.054	0.482	0.714	0.532	0.638
CAM-Conv5[a]	$64 \times 32 \times 32$	0.570	0.110	0.555	0.733	0.602	0.647
GradCAM-Conv1-lyr0[a]	$256 \times 128 \times 128$	0.570	0.110	0.555	0.733	0.602	0.647
GradCAM-Conv2-lyr3[a]	$256 \times 128 \times 128$	0.582	0.132	0.618	0.737	0.657	0.633
GradCAM-Conv3-lyr7[a]	$128 \times 64 \times 64$	0.589	0.142	0.600	0.742	0.642	0.649

[a]refers to occlusion of heat map using top x b-scans and depth rows.

score respectively. We should note that CAM-3D-CNN has shown to outperform the conventional machine learning with an increase of 5% in the AUC measure [9].

Figure 1 visualizes grad-CAM heat maps for two convolutional layers: conv#1 and conv#4 for healthy and glaucoma cases. It is clear from the table that the last/deeper convolutional layers yield general and global important regions across all cubes, while the first convolutional layers give more detailed highlights which are comparable to the segmentation of retinal layers. The field of view of the deeper layers is larger than the more superficial layers, but the size of the heat maps are smaller in the deeper layers. This contributes to the generation of heat maps highlight larger swathes of the retina, but also lacks detail. For example, the heat map size for conv#1 is $256 \times 128 \times 128$ while it is $16 \times 8 \times 8$ for conv#6. The 2nd column in Table 2 shows the heat map size for each convolutional layer. In the heat maps generated from conv#4 we see that the optic disc region is highlighted, which is a region known to be affected by glaucoma.

3.2   Occlusion Experiment

To quantitatively assess different grad-CAM visualization results, we calculated the performance measure drops of gradCAM-3D-CNN model using the occluded set resulted from each convolutional layer. We also computed the occluded test set using CAM heat map generated from CAM-3D-CNN model. In total, 7 different occluded sets were generated and the corresponding performance measures drops are reported in Table 2. From the table, the highest drop in accuracy is 40%, achieved by gradCAM-conv#6 layer, followed by gradCAM-conv#4, gradCAM-conv#5, CAM, gradCAM-conv#1, gradCAM-conv#2, and

Table 3. Occlusion results for different grad-CAM variants using Conv#6 (AUC of original model is 0.973)

	Accuracy	MCC	Recall	Precision	F1-Score	AUC
Grads[a]	0.520	0.032	0.459	0.705	0.509	0.596
Relu-grads[a]	0.526	0.041	0.468	0.709	0.519	0.636
Guided-grads[a]	0.529	0.045	0.473	0.711	0.523	0.627
Grads-pos[a]	0.529	0.045	0.473	0.711	0.523	0.628
Grads-abs[a]	0.531	0.050	0.477	0.713	0.528	0.616
Grads-neg[a]	0.534	0.054	0.482	0.714	0.532	0.610

[a]refers to occlusion of heat map using top x b-scans and depth rows.

gradCAM-conv#3 with accuracy drop of 39%, 39%, 35%, 35%, 34% and 33% in order. Further, heat map of gradCAM-conv#6 was also used to occlude the least important decision region, where the performance drop was only 4%, 3% and 4% in the AUC, accuracy and F1-score measures respectively. This confirms the effectiveness of grad-CAM for highlighting important decision regions.

Different variants of grad-CAM were implemented to enhance the heat maps by removing noisy gradients by either back propagating only positive gradients (relu-grads) [17] or positive gradients and positive input (guided-grads) [15]. Table 3 shows the impact of occlusion using different grad-CAM variants on the performance of the proposed model. We also investigated the impact of using different gradient modifiers, such as positive, negative, and absolute gradients, on the classification performance of the model. For example, in the case of absolute gradients we used the absolute values of the feature map gradients to compute the heat map. Table 3 shows that occlusion using grad-CAM without any modifier results in the highest drop in the performance.

4   Conclusion and Future Work

We present an end-to-end 3D CNN classification model that is able to effectively distinguish between healthy and glaucoma cases using 3D raw volumes. This approach improves on the accuracy of previously proposed methods [9], but also used an input that was double the size. This allowed for better CAMs to be generated using grad-CAM, which highlighted important regions of the retina. Further, grad-CAM heat maps were analyzed and quantitatively validated using the occlusion assessment method. In particular, the occlusion assessment method confirmed the effectiveness of grad-CAM in highlighting crucial decision regions. In the future, we will improve the evaluation using a cross-validation study, as well as extend this study to include other ocular diseases. We also plan to train the DL model on the important sub-volumes guided by grad-CAM results. We will also study the effect of fusing grad-CAM for different convolutional layers.

References

1. An, G., et al.: Glaucoma diagnosis with machine learning based on optical coherence tomography and color fundus images. J. Healthc. Eng. **2019**, 1–9 (2019)
2. Asaoka, R., et al.: Using deep learning and transfer learning to accurately diagnose early-onset glaucoma from macular optical coherence tomography images. Am. J. Ophthalmol. **198**, 136–145 (2019)
3. Davis, B.M., Crawley, L., Pahlitzsch, M., Javaid, F., Cordeiro, M.F.: Glaucoma: the retina and beyond. Acta Neuropathol. **132**(6), 807–826 (2016)
4. Flaxman, S.R., et al.: Global causes of blindness and distance vision impairment 1990–2020: a systematic review and meta-analysis. Lancet Glob. Health **5**(12), e1221–e1234 (2017)
5. Fujimoto, J., Swanson, E.: The development, commercialization, and impact of optical coherence tomography. Invest. Ophthalmol. Vis. Sci. **57**(9), OCT1–OCT13 (2016)
6. Glorot, X., Bordes, A., Bengio, Y.: Deep sparse rectifier neural networks. In: Proceedings of the Fourteenth International Conference on Artificial Intelligence and Statistics, pp. 315–323 (2011)
7. Gulli, A., Pal, S.: Deep Learning with Keras. Packt Publishing Ltd, Birmingham (2017)
8. Ioffe, S., Szegedy, C.: Batch normalization: accelerating deep network training by reducing internal covariate shift. arXiv preprint arXiv:1502.03167 (2015)
9. Maetschke, S., Antony, B., Ishikawa, H., Garvani, R.: A feature agnostic approach for glaucoma detection in OCT volumes. arXiv preprint arXiv:1807.04855 (2018)
10. Maetschke, S., Tennakoon, R., Vecchiola, C., Garnavi, R.: Nuts-flow/ml: data preprocessing for deep learning. arXiv preprint arXiv:1708.06046 (2017)
11. Muhammad, H., et al.: Hybrid deep learning on single wide-field optical coherence tomography scans accurately classifies glaucoma suspects. J. Glaucoma **26**(12), 1086–1094 (2017)
12. Razzak, M.I., Naz, S., Zaib, A.: Deep learning for medical image processing: overview, challenges and the future. In: Dey, N., Ashour, A.S., Borra, S. (eds.) Classification in BioApps. LNCVB, vol. 26, pp. 323–350. Springer, Cham (2018). https://doi.org/10.1007/978-3-319-65981-7_12
13. Selvaraju, R.R., Cogswell, M., Das, A., Vedantam, R., Parikh, D., Batra, D.: Grad-CAM: visual explanations from deep networks via gradient-based localization. In: Proceedings of the IEEE International Conference on Computer Vision, pp. 618–626 (2017)
14. Simonyan, K., Zisserman, A.: Very deep convolutional networks for large-scale image recognition. arXiv preprint arXiv:1409.1556 (2014)
15. Springenberg, J.T., Dosovitskiy, A., Brox, T., Riedmiller, M.: Striving for simplicity: the all convolutional net. arXiv preprint arXiv:1412.6806 (2014)
16. Wang, J., et al.: SD Net: joint segmentation and diagnosis revealing the diagnostic significance of using entire RNFL thickness in glaucoma. In: Conference on Medical Imaging with Deep Learning (MIDL) (2018)
17. Zeiler, M.D., Fergus, R.: Visualizing and understanding convolutional networks. In: Fleet, D., Pajdla, T., Schiele, B., Tuytelaars, T. (eds.) ECCV 2014. LNCS, vol. 8689, pp. 818–833. Springer, Cham (2014). https://doi.org/10.1007/978-3-319-10590-1_53
18. Zhou, B., Khosla, A., Lapedriza, A., Oliva, A., Torralba, A.: Learning deep features for discriminative localization. In: Proceedings of the IEEE Conference on Computer Vision and Pattern Recognition, pp. 2921–2929 (2016)

Semi-supervised Adversarial Learning for Diabetic Retinopathy Screening

Sijie Liu, Jingmin Xin$^{(\boxtimes)}$, Jiayi Wu, and Peiwen Shi

Institute of Artificial Intelligence and Robotics, Xi'an Jiaotong University,
Xi'an 710049, China
jxin@mail.xjtu.edu.cn

Abstract. It is well known that in medical image analysis, only a small number of high-quality labeled images can be often obtained from a large number of medical images due to the requirement of expert knowledge and intensive labor work. Therefore, we propose a novel semi-supervised adversarial learning framework (SSALF) for diabetic retinopathy (DR) screening of color fundus images. Specifically, our proposed framework consists of two subnetworks, an extended network and a discriminator. The extended network is obtained by extending a common classification network with a generator used for unsupervised image reconstruction. Thus, the extended network can utilize some labeled and lots of unlabeled fundus images. Then the discriminator is attached to the generator of the extended network to judge whether a reconstructed image is real or fake, introducing adversarial learning into the whole framework. Our framework achieves promising utility and generalization on the datasets of EyePACS and Messidor in a semi-supervised setting: we use some labeled and lots of unlabeled fundus images to train our framework. And we also investigate the effects of image reconstruction and adversarial learning on our framework by implementing ablation experiments.

1 Introduction

In many countries, diabetic retinopathy (DR) is the most common cause of blindness in adults. Fortunately, early diagnosis and timely treatment can effectively prevent the occurrence of blindness. With the development of color fundus photography, experienced ophthalmologists can observe various DR lesions in fundus images, rate the severity of DR, and decide corresponding treatments. To reduce the burden of ophthalmologists, various automatic DR screening methods [1] have been proposed. Recently, deep learning has become a leading methodology for medical image analysis and also has achieved promising performance [4] in DR screening. As we all know, a superior deep neural network usually involves large numbers of medical images with corresponding high-quality annotations. However, the process of obtaining these annotations not only is time-consuming, but also requires large amounts of expert knowledge. Hence, it is a challenging task to use a small number of labeled fundus images to achieve superior performance of DR screening.

© Springer Nature Switzerland AG 2019
H. Fu et al. (Eds.): OMIA 2019, LNCS 11855, pp. 60–68, 2019.
https://doi.org/10.1007/978-3-030-32956-3_8

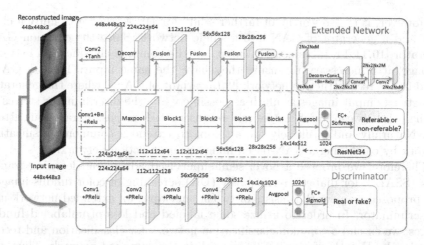

Fig. 1. Schematic of our semi-supervised adversarial learning framework (SSALF) for DR screening. Components of the extended network and the discriminator are in the purple and green solid boxes. ResNet34 and 'Fusion' are also showed in the red and yellow dotted boxes, where **'Block'** represents multiple stacked residual modules. (Color figure online)

Many works have been conducted to address the relative tasks [10,12,14,18, 19]. To reduce the number of images that need to be labeled, Yang et al. [18] exploited deep active learning to select the most effective medical images to be labeled. To utilize some labeled and lots of unlabeled images, Ladder network [12] and SWWAE [19] were proposed to simultaneously minimize the sum of the classification term and the reconstruction term in a semi-supervised setting. As the generative adversarial network (GAN) [3] becomes a research hotspot in semi-supervised and unsupervised learning, many researchers [6,10,14] proposed the GAN-based classification networks, such as ImprovedGAN [14]. In these networks, researchers unified a discriminator of GAN and a classifier into a single network. The new discriminator could predict N+1 classes, where N means categories of medical images, and 1 means whether a medical image is real or fake. Thus, during the training phase, these networks were trained with labeled medical images to predict N classes, and were trained with unlabeled and generated medical images to judge whether a medical image is real or fake. Lecouat et al. [8] proposed a patch-based semi-supervised classification approach to recognize abnormal fundus images, which was based on ImprovedGAN. TripleGAN [2] proposed a tripartite adversarial model, including three separated networks: a classifier, a generator and a discriminator. In fact, TripleGAN divided a discriminator of ImprovedGAN into two parts, the discriminator of TripleGAN and the classifier of TripleGAN, while adding an adversarial mechanism between the two parts. In these mentioned GAN-based methods, all the relationships between a generator and a classifier/a discriminator were cascading.

However, it is also worthy of further study for DR screening to unify a classifier and a generator of GAN into a single network. Since the common GAN generator [10,11,14] takes noise as input but the common classification network a image, it is necessary to ensure they have the same input. Thus, a GAN-based image reconstruction network [9] (a variant of GAN), where the generator reconstructs input images as much as possible while the discriminator strives to distinguish between input images and reconstructed images, attracts our attention. Naturally and intuitively, we attempt to extend a common classification network by combining a GAN-based image reconstruction network.

Therefore, we propose a novel semi-supervised adversarial learning framework (SSALF) for diabetic retinopathy (DR) screening of color fundus images. Our proposed framework consists of two subnetwork, an extended network and a discriminator. In order to utilize some labeled and lots of unlabeled fundus images, we extend a common classification network for classification and reconstruction by U-net's "fusion" [13], and call it the extended network. Thus, the extended network comprises two components, a classifier for supervised classification and a generator for unsupervised image reconstruction. Then like a common GAN, we attach the discriminator to the generator of the extended network to judge whether a reconstructed image is real or fake, which introduces adversarial learning into the whole framework. In summary, our contribution are as follows: (i) We propose a novel semi-supervised adversarial learning framework for DR screening, extending a common classification network by combining a GAN-based image reconstruction network. (ii) We also propose an appropriate training strategy to effectively and efficiently train our framework. (iii) Our framework achieves promising utility and generalization on the datasets of Eye-PACS and Messidor in a semi-supervised setting: we use some labeled and lots of unlabeled fundus images to train our framework. We also investigate the effects of image reconstruction and adversarial learning on our framework by implementing ablation experiments.

2   Methods

2.1   Common Networks for Classification

With the rapid development of deep convolution neural networks (DCNNs), some common classification networks [5,7,15,16] were proposed successively. Nowadays, many works [17] modified these common networks according to specific medical image analysis tasks, and achieved promising results. Thus, considering convergence speed and memory overhead, we exploit ResNet34 [5] (denoted as C) as the base model, and use the binary cross entropy loss as a loss function:

$$\mathcal{L}_{CLS}^{sup}(C) = -(y \log C(\mathcal{X}) + (1-y) \log(1 - C(\mathcal{X})), \tag{1}$$

where y is a class label of an input image $\mathcal{X}$ from a supervised subset.

2.2   Semi-supervised Adversarial Learning Framework

Unlike the aforementioned GAN-based methods [2,6,8,10,14], our framework focuses on unifying a classifier and a generator of GAN into a single network, instead of dividing them into two cascading models. Our framework consists of two subnetworks, i.e., (1) an extended network for classification and reconstruction, and (2) a discriminator for introducing adversarial learning. For the extended network, we use ResNet34 as a backbone architecture (which is served as a supervised classifier (donated as C)) and extend it with an unsupervised generator (donated as G). Specifically, we use ResNet34 as our classifier, but the fully connected layer of ResNet34 is modified to output two-dimension values. Then we use the U-net's "fusion" (See Fig. 1) and deconvolution to upsample from the last convolutional layer of our classifier until the output size is consistent with the input image, which constructs our generator and makes it can combine low-level and high-level features to reconstruct input images. In our point of view, the generator not only acts as a regularizer [19], but also forces the classifier to focus on abstract invariant features on the higher level [12] by utilizing the "fusion". For the discriminator (donated as D), following the rules in [11], we design a simple seven-layer DCNN. We aim to add a regularization item, which can learn the distribution of real fundus images, to the extended network by introducing adversarial learning. Our framework schematic is depicted in Fig. 1, and is theoretically easy to deploy to the other aforementioned common networks.

Our pipeline is that the extended network outputs results of classification and reconstruction simultaneously, and that then the discriminator determines whether a fundus image is reconstructed by the extended network or not. Therefore, the classification and the GAN-based image reconstruction are unified into a single framework.

The total loss of the extended network is formulated as:

$$\mathcal{L}_{EXT}^{semi}(C,G) = \mathcal{L}_{CLS}^{sup}(C) + \mu(\mathcal{L}_{MSE}^{unsup}(G) + \lambda\mathcal{L}_{AD\_G}^{unsup}(G)), \qquad (2)$$

where

$$\mathcal{L}_{MSE}^{unsup}(G) = \frac{1}{WH}\sum_{x=1}^{W}\sum_{y=1}^{H}(I_{x,y} - G(I_{x,y}))^2, \qquad (3)$$

$$\mathcal{L}_{AD\_G}^{unsup}(G) = -\log D(G(\mathcal{X})), \qquad (4)$$

where $\mathcal{L}_{MSE}^{unsup}(G)$ and $\mathcal{L}_{AD\_G}^{unsup}(G)$ indicate the image reconstruction loss of the generator and the adversarial loss of the generator in an unsupervised subset, respectively. $\mathcal{L}_{CLS}^{sup}(C)$ represents the binary cross entropy loss of the classifier in a supervised subset. μ and λ refer to the weighting coefficients of the unsupervised loss and the adversarial loss, respectively. And W and H denote the size of a input image $\mathcal{X}$ while $I_{x,y}$ indicates the image pixel value.

The adversarial loss of the discriminator is formulated as:

$$\mathcal{L}_{AD\_D}^{unsup}(D) = -(\log D(\mathcal{X}) + \log(1 - D(G(\mathcal{X})))). \qquad (5)$$

2.3 Appropriate Training Strategy

Training a deep neural network is also non-trivial. Therefore, we propose the following steps to effectively and efficiently train our framework.

1. We use weights of ResNet34 pre-trained on the ILSVRC to initialize the extended network [17], and train it only for image reconstruction.

2. We use the weights trained in step 1 to reinitialize the extended network.

3. We fix the extended network, and then train the discriminator once with minimization of $\mathcal{L}_{AD\_D}^{unsup}(D)$.

4. We fix the discriminator, and then train the extended network once with minimization of $\mathcal{L}_{EXT}^{semi}(C, G)$.

5. We iterate the step 3 and 4 until the extended network converges.

Among the above steps, step 1 has been demonstrated to be quite effective in [17], and step 2 is extremely crucial for semi-supervised DR screening, which will be demonstrated in Sect. 3.2. Pytorch[1] is adopted to implement our proposed framework. Scaling radius, random crop, random translation, random rotation and random flip are applied to preprocess and augment our dataset. Besides, all the fundus images are resized to $448 \times 448 \times 3$. Our framework is trained on a Nvidia GTX 1080Ti of 11 GB memory with a batch size of 16. The Nesterov SGD algorithm with an initial learning rate of 1e−3, a momentum of 0.9 and a weight decay of 5e−4 is used to optimize the extended network and the discriminator during the training. μ is set as 50 initially and will be reduced later in order to keep the ratio of losses between the classifier and the generator more than 4:1. And λ is set as 3e-4.

3 Experiments

3.1 Dataset Description

Our framework is evaluated on two publicly available datasets: the dataset of 'Kaggle Diabetic Retinopathy Detection' (EyePACS)[2] and the Messidor dataset[3].

The EyePACS dataset contains 35,126 training images with graded labels and 53,576 test images without graded labels. The presence of the diabetic retinopathy in each image has been graded by a clinician into one of the five stages: no DR, mild, moderate, severe, and proliferative DR. Here we only focus on the non-referable DR stage (including the no DR stage and the mild stage) and the referable DR stage (including the moderate stage, the severe stage, and proliferative DR stage). We divide the training images into three subsets: kaggle-train (the first 21,076 images), kaggle-val (the middle 7026 images), and kaggle-test (the last 7026 images). In our semi-supervised setting, we randomly select 500,

---

[1] https://github.com/pytorch/pytorch.

[2] https://www.kaggle.com/c/diabetic-retinopathy-detection/data.

[3] http://www.adcis.net/en/Download-Third-Party/Messidor.html.

Table 1. AUC of different methods on the EyePACS dataset.

Method	500	1000	2000	3000
ResNet34	0.773	0.833	0.869	0.886
ImprovedGAN	**0.806**	**0.858**	0.876	0.890
SSALF (ours)	0.800	0.854	**0.883**	**0.900**

Table 2. Ablation experiments on the EyePACS dataset.

Method	500		1000		2000		3000	
	AUC	SSIM	AUC	SSIM	AUC	SSIM	AUC	SSIM
ResNet34	0.773	-	0.833	-	0.869	-	0.886	-
ResNet34+Rec*	0.751	0.748	0.828	0.841	0.871	0.892	0.887	0.905
ResNet34+Rec	0.791	0.891	0.847	0.892	0.879	0.929	0.895	**0.939**
SSALF (ours)	**0.800**	**0.893**	**0.854**	**0.910**	**0.883**	**0.939**	**0.900**	0.932

1000, 2000 and 3000 images from the kaggle-train as supervised subsets respectively. Meanwhile, we only use the entire kaggle-train as a unsupervised subset. These subsets are balanced by oversampling (random crop).

The Messidor dataset contains 1,200 color fundus images. Different from the EyePACS dataset, the Messidor dataset divides all the images into four stages. Similarly, we can obtain 699 non-referable fundus images and 501 referable fundus images from this dataset. Here we use the whole Messidor dataset as an independent dataset for test.

3.2 Experiment Results

We perform semi-supervised experiments on the datasets of EyePACS and Messidor. The area under the receiver operating curve (AUC) and the structural similarity index (SSIM) are used to quantify the performance of the classification and the image reconstruction, respectively.

EyePACS: To evaluate the performance of our proposed framework, we compare our framework with ResNet34 and ImprovedGAN [14], as shown in Table 1. To make a fair comparison, we adopt ResNet34 as the discriminator of ImprovedGAN. For the generator of ImprovedGAN, we use 200-dimension vectors as input and add several deconvolutional layers to the original version [14] in order to generate $448 \times 448 \times 3$ fundus images. It is observed in Table 1 that our SSALF trained with 500 or 1000 labeld fundus images can achieve comparable AUCs with ImprovedGAN while with the increase of labeld fundus images, our SSALF can achieve more improvements than ImprovedGAN. Furthermore, in the case of 2000 or 3000 labeled fundus images, ImprovedGAN only achieves a little improvement compared to ResNet34 while our SSALF doesn't show significant gain reduction.

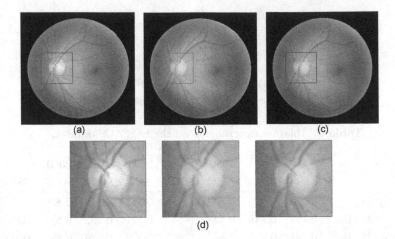

Fig. 2. Fake images generated from (a) an input image of the kaggle-test through (b) ResNet34+Rec and (c) our SSALF, where (d) optic disks are selected for comparison.

Table 3. AUC of different methods on the Messidor dataset.

Method	500	1000	2000	3000
ResNet34	0.817	0.892	0.923	0.932
ImprovedGAN	**0.907**	**0.922**	0.934	0.941
SSALF (ours)	0.877	0.910	**0.936**	**0.945**

To investigate the effects of the image reconstruction and the adversarial learning respectively, we conduct several ablation experiments, as shown in Table 2. ResNet34+Rec (ResNet34+Rec*) indicates the extended network with (without) the initialization in the aforementioned step 2. We can find in Table 2 that ResNet34+Rec* can only achieve the comparable results with ResNet34. (1) This shows that without good initialization in the aforementioned step 2, the generator can't provide good regularization for the classifier during the training. Particularly noting, the adversarial learning has no relationship with improving SSIM, which is also pointed out in [9]. Figure 2 displays the fake images generated from the kaggle-test by using ResNet34+Rec and our SSALF. A closer look reveals our SSALF produces thinner but clearer texture, especially texture in the optic disk. In Table 2 we can also find that with different numbers of labeled fundus images, ResNet34+Rec can achieve better AUC than ResNet34 while our SSALF achieves the best AUC. (2) This shows combining the image reconstruction can indeed improve the performance of DR screening dramatically, while introducing the adversarial learning can further enhance the performance.

Messidor: In order to demonstrate the generalization ability of our framework, we also evaluate it on the Messidor dataset, but only use this dataset for testing.

Results are shown in Table 3. It is observed that all the results from the Messidor dataset keep the same trend with those from the EyePACS dataset, and that our SSALF can even achieve an AUC of 0.945. This shows that our framework has good generalization ability.

4 Conclusions

In this paper, we propose a novel semi-supervised adversarial learning framework for diabetic retinopathy screening of color fundus images, and an appropriate training strategy. Experiment results on the datasets of EyePACS and Messidor show that our framework can achieve comparable or better utility and generalization than ImprovedGAN. Our ablation experiments show that combining the image reconstruction can indeed improve the performance dramatically, while introducing the adversarial learning can further enhance the performance.

Acknowledgements. This work was supported in part by the Programme of Introducing Talents of Discipline to University: B13043, and the National Key Research and Development Program of China under grant 2017YFA0700800.

References

1. Abràmoff, M.D., et al.: Automated early detection of diabetic retinopathy. Ophthalmology **117**(6), 1147–1154 (2010)
2. Chongxuan, L., Xu, T., Zhu, J., Zhang, B.: Triple generative adversarial nets. In: NIPS, pp. 4088–4098 (2017)
3. Goodfellow, I., et al.: Generative adversarial nets. In: NIPS. pp. 2672–2680 (2014)
4. Gulshan, V., et al.: Development and validation of a deep learning algorithm for detection of diabetic retinopathy in retinal fundus photographs. JAMA **316**(22), 2402–2410 (2016)
5. He, K., Zhang, X., Ren, S., Sun, J.: Deep residual learning for image recognition. In: CVPR, pp. 770–778 (2016)
6. Hu, B., et al.: Unsupervised learning for cell-level visual representation in histopathology images with generative adversarial networks. IEEE J. Biomed. Health Inform. **23**(3), 1316–1328 (2018)
7. Krizhevsky, A., Sutskever, I., Hinton, G.E.: ImageNet classification with deep convolutional neural networks. In: NIPS, pp. 1097–1105 (2012)
8. Lecouat, B., et al.: Semi-supervised deep learning for abnormality classification in retinal images. In: Machine Learning for Health (ML4H) Workshop at NeurIPS (2018)
9. Ledig, C., et al.: Photo-realistic single image super-resolution using a generative adversarial network. In: CVPR, pp. 4681–4690 (2017)
10. Madani, A., Moradi, M., Karargyris, A., Syeda-Mahmood, T.: Semi-supervised learning with generative adversarial networks for chest x-ray classification with ability of data domain adaptation. In: ISBI 2018, pp. 1038–1042. IEEE (2018)
11. Radford, A., Metz, L., Chintala, S.: Unsupervised representation learning with deep convolutional generative adversarial networks. arXiv preprint arXiv:1511.06434 (2015)

12. Rasmus, A., Berglund, M., Honkala, M., Valpola, H., Raiko, T.: Semi-supervised learning with ladder networks. In: NIPS, pp. 3546–3554 (2015)
13. Ronneberger, O., Fischer, P., Brox, T.: U-Net: convolutional networks for biomedical image segmentation. In: Navab, N., Hornegger, J., Wells, W.M., Frangi, A.F. (eds.) MICCAI 2015. LNCS, vol. 9351, pp. 234–241. Springer, Cham (2015). https://doi.org/10.1007/978-3-319-24574-4_28
14. Salimans, T., Goodfellow, I., Zaremba, W., Cheung, V., Radford, A., Chen, X.: Improved techniques for training GANS. In: NIPS, pp. 2234–2242 (2016)
15. Simonyan, K., Zisserman, A.: Very deep convolutional networks for large-scale image recognition. arXiv preprint arXiv:1409.1556 (2014)
16. Szegedy, C., et al.: Going deeper with convolutions. In: CVPR, pp. 1–9 (2015)
17. Vo, H.H., Verma, A.: New deep neural nets for fine-grained diabetic retinopathy recognition on hybrid color space. In: 2016 IEEE International Symposium on Multimedia (ISM), pp. 209–215. IEEE (2016)
18. Yang, L., Zhang, Y., Chen, J., Zhang, S., Chen, D.Z.: Suggestive Annotation: a deep active learning framework for biomedical image segmentation. In: Descoteaux, M., Maier-Hein, L., Franz, A., Jannin, P., Collins, D.L., Duchesne, S. (eds.) MICCAI 2017. LNCS, vol. 10435, pp. 399–407. Springer, Cham (2017). https://doi.org/10.1007/978-3-319-66179-7_46
19. Zhao, J., Mathieu, M., Goroshin, R., Lecun, Y.: Stacked what-where auto-encoders. arXiv preprint arXiv:1506.02351 (2015)

Shape Decomposition of Foveal Pit Morphology Using Scan Geometry Corrected OCT

Min Chen[1]([✉]), James C. Gee[1], Jessica I. W. Morgan[2,3], and Geoffrey K. Aguirre[4]

[1] Department of Radiology, University of Pennsylvania, Philadelphia, PA 19104, USA
minchen1@upenn.edu
[2] Scheie Eye Institute, Department of Ophthalmology, University of Pennsylvania, Philadelphia, PA 19104, USA
[3] Center for Advanced Retinal and Ocular Therapeutics, University of Pennsylvania, Philadelphia, PA 19104, USA
[4] Department of Neurology, University of Pennsylvania, Philadelphia, PA 19104, USA

Abstract. The fovea is an important structure that allows for the high acuity at the center of our visual system. While the fovea has been well studied, the role of the foveal pit in the human retina is still largely unknown. In this study we analyze the shape morphology of the foveal pit using a statistical shape model to find the principal shape variations in a cohort of 50 healthy subjects. Our analysis includes the use of scan geometry correction to reduce the error from inherent distortions in OCT images, and a method for aligning foveal pit surfaces to remove translational and rotational variability between the subjects. Our results show that foveal pit morphology can be represented using less than five principal modes of variation. And we find that the shape variations discovered through our analysis are closely related to the main metrics (depth and diameter) used to study the foveal pit in current literature. Lastly, we evaluated the relationship between the first principal mode of variation in the cohort and the axial length from each subject. Our findings showed a modest inverse relationship between axial length and foveal pit depth that can be confirmed independently by existing studies.

Keywords: Shape analysis · Distortion correction · Retina · Fovea · OCT

1 Introduction

The fovea is a critical structure in the retina that consists of tightly packed cone photoreceptor cells, which allows for the high acuity at the center of our visual system. In humans, the fovea is characterized by a depression (*foveal pit*) where the retinal ganglion and bipolar cells are displaced. While studies have

© Springer Nature Switzerland AG 2019
H. Fu et al. (Eds.): OMIA 2019, LNCS 11855, pp. 69–76, 2019.
https://doi.org/10.1007/978-3-030-32956-3_9

characterized the variability of the fovea [1], the role of the foveal pit in our visual system is still unclear. Recent studies have shown that subjects who lack a foveal pit can still maintain cone specialization and normal acuity [2,3]. Given the importance of the fovea, studying the morphology and structure of the foveal pit can lead to a better understanding of diseases that affect our central vision.

Foveal pit morphology has been studied in relationship with other retinal measures such as thickness [4], foveal avascular zone [5], and visual acuity [6]. Notably, Wilk et al. [3] presented a study using multi-modal imaging to evaluate the relationship between foveal pit morphology from optical coherence tomography (OCT) imaging and cone density measures from adaptive optics scanning laser ophthalmoscopy (AOSLO) from subjects with albinism, showing considerable variation between the two measures. One limitation of these existing studies is their reliance on the evaluation of summary measurements (pit depth, diameter and slope) [7] or parametric models [8] of the foveal pit. Such analysis are restricted by the location of the measurement or the fit of the model, and do not observe the 3D local or spatial relationship of the foveal pit morphology across a population.

OCT offers in-vivo 3D imaging of the retina, and is currently the most effective imaging modality for observing the shape morphology of the foveal pit. However, one challenge with studying shape morphology in OCT is the presence of scan geometry distortion due to the instrument acquiring the images in a fan-beam pattern, but representing it as a rectangular grid [9,10]. Currently, such distortions are not corrected before analyzing the shape morphology of the foveal pit, which can have significant impact on the shape analysis.

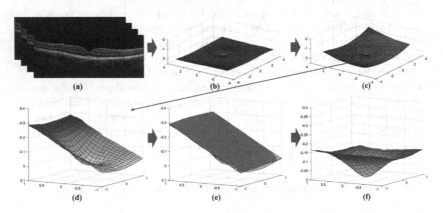

Fig. 1. Diagram showing our processing pipeline to extract the foveal pit surface from each OCT image. (a) shows the raw OCT image and delineations found for the retinal layers. (b) shows the inner limiting membrane surface prior to scan geometry correction and (c) shows the surface after the distortion correction. (d) is an extracted 1 mm by 1 mm region surrounding the fovea. (e) shows a planar fit to the retinal surface, and (f) shows the foveal pit surface after using the fitted plane to correct for rotations in the image.

The goal of this work is to improve on existing analysis of the foveal pit shape morphology through the use of a statistical shape model to decompose the principal modes of shape variation across a cohort of 50 subjects. In addition, we aim to improve the general reliability of foveal pit shape analysis by correcting the OCT geometry of our scans [10] and addressing the translational and rotational variability of the foveal pit surfaces between subjects.

2   Method

2.1   OCT Geometry Correction and Surface Construction

It is well established that OCT images do not represent a Euclidean space despite their presentation of the data in a rectangular pattern [9,10]. The individual columns (A-scans) of an OCT are path measurements that traverse a fan-beam pattern that spreads out from a central nodal point located in the eye. The distance between the A-scans are in units of degrees, and in actual Euclidean space the OCT image would look similar to an ultrasound image, where the top of the image is more narrow than the bottom. This misrepresentation of the OCT scan geometry results in a distortion of the retina morphology that can impact our shape analysis. To address this distortion, we use a digital model [11] of each of our OCT images to approximate the location of the A-scan nodal point. We then correct the scan geometry (Fig. 1c) of the OCT using an established model for distortion correction [10] and interpolating the OCT within this corrected space. From the OCT we segment the inner limiting membrane and extract a 1 mm by 1 mm region surrounding the fovea to get a surface representation of the foveal pit (Fig. 1d).

2.2   Foveal Pit Alignment

A statistical shape model [12] requires that the images from every subject are aligned into a common space. This allows us to establish correspondences and remove incidental variation in the data (such as global movement) that should not be included into the analysis. For our study, we perform a two-step rigid alignment of the foveal pit surfaces. First, we automatically find the deepest point in each foveal pit as a landmark, which we refer to as the *fovea center*. Each surface is translated such that the fovea center is moved to the origin of the coordinate system. This ensures that every foveal pit surface is centered and has a coherent point of reference. However, after translation, the fovea pit may be tilted in different orientations due to the positioning of the retina and OCT scanning angle. We address this by first fitting a plane to each retinal surface (Fig. 1e). The foveal pit is then rotated around the fovea center until the retinal plane is parallel to the X-Y plane of the coordinate system (Fig. 1f).

2.3 Principal Shape Decomposition

Given a surface of the foveal pit from each subject, we represent the surface data as a vector:

$$\mathbf{X} = [x_1 \ y_1 \ z_1 \ x_2 \ y_2 \ z_2 \ ... \ x_V \ y_V \ z_V], \tag{1}$$

where (x_v, y_v, z_v) for $v \in [1, 2...V]$ are the 3D positions of each of the V vertices in the surface. Each surface is resampled such that they have the same number of vertices and the x and y coordinates fall on a common grid. Given N subjects, we can then stack the surfaces from each subject into a single data matrix:

$$\mathbf{D} = [\mathbf{X}_1 \ \mathbf{X}_2 \ \mathbf{X}_3 \ ... \ \mathbf{X}_N]^T, \tag{2}$$

where each row of the matrix represents the surface data from each subject. Using $\mathbf{D}$ we can perform a shape decomposition of the foveal pit using principal component analysis [13] to find the principal modes of variation in the data. To do this, we first subtract the mean from the data to prevent the principal components from being directed by the global bias in the data. This is calculated by evaluating the mean across the surfaces

$$\overline{\mathbf{X}} = \sum_{n=1}^{N} X_n, \tag{3}$$

and then subtracting it from each surface to create a new data matrix

$$\hat{\mathbf{D}} = [(\mathbf{X}_1 - \overline{\mathbf{X}}) \ (\mathbf{X}_2 - \overline{\mathbf{X}}) \ (\mathbf{X}_3 - \overline{\mathbf{X}}) \ ... \ (\mathbf{X}_N - \overline{\mathbf{X}})]^T, \tag{4}$$

which has zero mean. Singular value decomposition is then applied to $\hat{\mathbf{D}}$ to find the linear relationship:

$$\mathbf{T} = \hat{\mathbf{D}}\mathbf{W}, \tag{5}$$

where the columns of $\mathbf{W}$ are orthogonal unit vectors $\{\mathbf{w}_1 ... \mathbf{w}_V\}$ of size V that describe the principal modes of variation (also known as the principal components [PC]) in the data, and $\mathbf{T}$ is a matrix where each row is the projection of the data from each subject into the PC space. Thus, each element of $t_{n,v} \in \mathbf{T}$ is the *PC score* of subject n with regard to vth PC. In the next section we show how each PC can be added to the mean shape of the population to view the different types of shape variation in the data. In addition, we will demonstrate how the PC scores of the subjects can be used as a quantitative measure that can be compared to external metrics of the retina, such as axial length. This serves as a pilot study for future analysis where we aim to use this technique to establish relationships between retinal shape and clinical measures of disease.

3 Evaluation and Results

3.1 Data

We analyzed 50 macular OCT images collected from healthy subjects using a Spectralis (Heidelberg Engineering, Dossenheim, Germany) scanner. Each image covered a 30° by 20° field of view centered on the fovea and had an associated segmentation of the ILM layer delineated automatically using OCTExplorer [14]. The foveal pit was extracted from each OCT image and aligned to the atlas space as described in Sect. 2 (and shown in Fig. 1).

3.2 Shape Decomposition of the Foveal Pit

Using the shape decomposition described before, we calculated the primary modes of shape variation of the foveal pit in our cohort. Figure 2a shows the percent of shape variation that each PC accounts for across the total shape variation in the data. We see that the first PC covers over 80% of the total shape variation, and the first 5 PCs together cover over 97% of the shape variation. Figure 3 shows a visualization of each of the first 5 PCs as they range from -2 to $+2$ standard deviation of the population relative to the mean shape. We observed that the first PC represented the variability in the depth of the fovea pit. The second and third PCs represented global tilt (in orthogonal directions) of the foveal pit. The fourth PC represented the diameter of the pit. And lastly, the fifth PC represented slight changes in the regions surrounding the foveal pit.

3.3 Correlation with Axial Length

One strength of performing shape decomposition is the ability to compare the PC shape scores with relevant biometric and clinical measures. Such analyses allow us to establish relationships between the shape morphology observed in the images and measures of anatomy or disease. Understanding these relationships can help us better understand the eye and also provide potential biomarkers or predictors of disease progression. To demonstrate this type of analysis, we correlate each PC1 score from our shape decomposition with the axial length of each individual's eye. Figure 2b shows a scatter plot of this relationship. We observe from this analysis that there is a modest ($r = 0.51$) Pearson's correlation between the two measures.

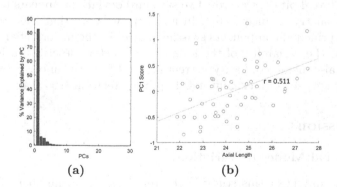

     (a)   (b)

Fig. 2. (a) shows the percentage of the total shape variability accounted for by each principal mode of variation (PC) in our shape analysis. (b) shows a scatter plot relating each subject's PC1 score with their axial length.

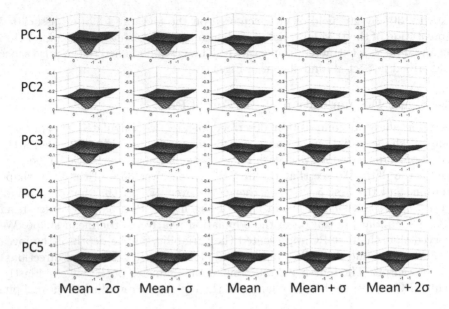

Fig. 3. Visualization of the shape variability represented by the five largest principal modes of variation (PC) found in our shape analysis. For each PC, we show the mean foveal pit shape of the population ±2 standard deviations (σ) of the shape variability observed in the population for that PC.

3.4 Impact of the Geometry Correction

To evaluate the role that the scan geometry correction had on the shape analysis, we repeated the principal shape decomposition on the foveal pit surfaces without first correcting for the distortion. Figure 4 shows a comparison of the first PC when using and not using the geometry correction. From the figure we can see that there is a significant change to the principal mode of variation. The total range of the foveal pit depth across two standard deviations appears to be larger when the geometry is uncorrected. In addition, we note that the order of the 3rd and 4th principal components switched places in the uncorrected case. This suggests that the variability of the foveal pit diameter (described originally by the 4th PC) also increased. Lastly, correlating PC1 from the uncorrected analysis with axial length showed a significant drop in correlation (r = 0.42).

4   Discussion

4.1   Principal Modes of Variation

One notable finding of this study is the relatively small number of principal components that is required to represent the shape variation in the foveal pit. PC1, which describes the depth of the foveal pit, covers over 80% of the shape variation. This is in line with existing analysis of the foveal pit [7,8], which relies

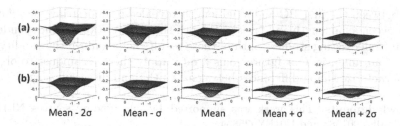

Fig. 4. Comparison of the first PC when the shape analysis was performed (a) with and (b) without OCT scan geometry correction. Shown is the mean foveal pit shape of the population ±2 two standard deviations (σ) of the PC variability observed in the population.

on explicit measurements or modeling of pit depth and diameter. It is reassuring to see that our unsupervised analysis was able to automatically find the same characteristics (PC1 and PC4) of the foveal pit that is deemed important by the community. However, the advantage of using our PCs over the existing summary measurements is that each PC covers the shape variation across the entire foveal pit. From Fig. 3 we see that the primary modes of shape variation are more complex than simply making a single measurement of depth or diameter. Our analysis shows other distinct variations that subtly changes the structure of the pit. This allows us to establish local spatial relationships in the foveal pit that are not covered by the common summary measurements used in the current literature.

4.2 Comparison to Existing Literature

In Fig. 2b, we found a modest positive relationship between each subject's score for the first principal mode of variation (PC1) and their axial length measurement. From Fig. 3 we see that PC1 has an inverse relationship with foveal pit depth (the pit becomes more shallow as PC1 increase). Thus, our shape analysis showed a morphological relationship where foveal pit depth decreases with increasing axial length. From existing literature, it is well established that axial length measurements increase with myopia [15]. Likewise, myopic eyes have been shown to have more shallow foveal pits [16]. Thus, this relationship confirms our finding that increased axial length is related to smaller foveal pit depths.

5 Conclusion

We performed a study of foveal pit morphology over a cohort of 50 health subjects using shape decomposition. Our analysis showed that the morphology of the foveal pit can be represented by as few as 5 principal modes of variation. The modes of variation we found were strongly related with metrics (pit depth and diameter) commonly used in existing analysis of the foveal pit. A correlative study between the first PC and axial length revealed a modest inverse relationship between axial length and foveal pit depth that we were able to independently confirm from existing studies. Both of these results help confirm the validity and value of our analysis. Our future goals are to apply these techniques to observe and characterize shape differences between healthy and disease cohorts.

Acknowledgments. This work was supported by our funding sources NEI/NIH grants P30EY001583 and U01EY025864.

References

1. Tick, S., et al.: Foveal shape and structure in a normal population. Invest. Ophthalmol. Vis. Sci. **52**(8), 5105–5110 (2011)
2. Marmor, M.F., Choi, S.S., Zawadzki, R.J., Werner, J.S.: Visual insignificance of the foveal pit: reassessment of foveal hypoplasia as fovea plana. Arch. Ophthalmol. **126**(7), 907–913 (2008)
3. Wilk, M.A., et al.: Relationship between foveal cone specialization and pit morphology in albinism. Invest. Ophthalmol. Vis. Sci. **55**(7), 4186–4198 (2014)
4. Wagner-Schuman, M., et al.: Race-and sex-related differences in retinal thickness and foveal pit morphology. Invest. Ophthalmol. Vis. Sci. **52**(1), 625–634 (2011)
5. Samara, W.A., et al.: Correlation of foveal avascular zone size with foveal morphology in normal eyes using optical coherence tomography angiography. Retina **35**(11), 2188–2195 (2015)
6. Williams, D.R.: Visual consequences of the foveal pit. Invest. Ophthalmol. Vis. Sci. **19**(6), 653–667 (1980)
7. Moore, B.A., Yoo, I., Tyrrell, L.P., Benes, B., Fernandez-Juricic, E.: FOVEA: a new program to standardize the measurement of foveal pit morphology. PeerJ **4**, e1785 (2016)
8. Dubis, A.M., McAllister, J.T., Carroll, J.: Reconstructing foveal pit morphology from optical coherence tomography imaging. Br. J. Ophthalmol. **93**(9), 1223–1227 (2009)
9. Westphal, V., Rollins, A., Radhakrishnan, S., Izatt, J.: Correction of geometric and refractive image distortions in optical coherence tomography applying Fermat's principle. Opt. Express **10**, 397–404 (2002)
10. Kuo, A.N., et al.: Correction of ocular shape in retinal optical coherence tomography and effect on current clinical measures. Am. J. Ophthalmol. **156**, 304–11 (2013)
11. Aguirre, G.K.: A model of the entrance pupil of the human eye. bioRxiv (2018)
12. Heimann, T., Meinzer, H.-P.: Statistical shape models for 3D medical image segmentation: a review. Med. Image Anal. **13**(4), 543–563 (2009)
13. Stegmann, M.B., Gomez, D.D.: A brief introduction to statistical shape analysis. Inf. Math. Model. **15**(11) (2002)
14. Garvin, M.K., Abràmoff, M.D., Wu, X., Russell, S.R., Burns, T.L., Sonka, M.: Automated 3-D intraretinal layer segmentation of macular spectral-domain optical coherence tomography images. IEEE Trans. Med. Imaging **28**, 1436–47 (2009)
15. Llorente, L., Barbero, S., Cano, D., Dorronsoro, C., Marcos, S.: Myopic versus hyperopic eyes: axial length, corneal shape and optical aberrations. J. Vis. **4**, 288–98 (2004)
16. Stroupe, R., Coletta, N.: Myopic versus hyperopic eyes: axial length, corneal shape and optical aberrations. Invest. Ophthalmol. Vis. Sci. **54**(15), 3610 (2013)

U-Net with Spatial Pyramid Pooling for Drusen Segmentation in Optical Coherence Tomography

Rhona Asgari[✉], Sebastian Waldstein, Ferdinand Schlanitz,
Magdalena Baratsits, Ursula Schmidt-Erfurth, and Hrvoje Bogunović

Christian Doppler Laboratory for Ophthalmic Image Analysis,
Department of Ophthalmology, Medical University of Vienna, Vienna, Austria
fatemeh.asgari@meduniwien.ac.at

Abstract. The presence of drusen is the main hallmark of early/ intermediate age-related macular degeneration (AMD). Therefore, automated drusen segmentation is an important step in image-guided management of AMD. There are two common approaches to drusen segmentation. In the first, the drusen are segmented directly as a binary classification task. In the second approach, the surrounding retinal layers (outer boundary retinal pigment epithelium (OBRPE) and Bruch's membrane (BM)) are segmented and the remaining space between these two layers is extracted as drusen. In this work, we extend the standard U-Net architecture with spatial pyramid pooling components to introduce global feature context. We apply the model to the task of segmenting drusen together with BM and OBRPE. The proposed network was trained and evaluated on a longitudinal OCT dataset of 425 scans from 38 patients with early/intermediate AMD. This preliminary study showed that the proposed network consistently outperformed the standard U-net model.

1 Introduction

Age-related macular degeneration (AMD) is a devastating retinal disease and a leading cause of blindness in the elderly population in the developed world [1]. The clinical hallmark and usually the first finding of AMD is the presence of waste deposits, called drusen. In the early stages, these drusen begin to accumulate in between two anatomical layers of the retina, the outer boundary retinal pigment epithelium (OBRPE) and the Bruch's membrane (BM). The drusen buildup and the consequent AMD progression to late stages are remarkably variable among affected individuals, resulting in its management being one of the biggest dilemmas in ophthalmology [2]. Currently, the patient scheduling frequency is primarily guided by the amount of drusen, which is subjectively assessed by drusen segmentation in optical coherence tomography (OCT). OCT is the state-of-the-art imaging modality for assessing the retina in AMD. This fast

© Springer Nature Switzerland AG 2019
H. Fu et al. (Eds.): OMIA 2019, LNCS 11855, pp. 77–85, 2019.
https://doi.org/10.1007/978-3-030-32956-3_10

and non-invasive acquisition technique allows to inspect the retina at a microme-
ter resolution, granting the possibility to study not only the retinal layers but also
several disease-related abnormalities, including drusen. Manual drusen segmentation is very time consuming, which creates a need for advanced medical image
computing methods that can measure distinct and pathognomonic changes in
drusen morphology in an accurate, objective and reproducible manner.

Related Work. In recent years, deep learning based and non deep learning
based methods were applied on this task [3–7]. Generally it has been shown that
deep learning based methods, namely convolutional neural networks (CNN),
outperform the previous cost-function based models [3,6,7]. In [3] a basic U-Net
is applied on drusen and layer segmentation. In [6] a combination of a CNN,
graph search based methods and standard classifier is introduced. In [7] a retina
layer segmentation task is tackled by a B-scan level CNN.

Drusen segmentation task can be tackled by segmenting the neighbouring
layers in the retina: BM and OBRPE. An alternative approach is to segment
drusen as an additional class. Our assumption is that this additional class will
not only provide more information about the layers adjacent to drusen class, but
will also help the network to characterize the appearance of both drusen and non-
pathological regions where OBRPE and BM overlap. The size of drusen varies,
meaning a given drusen could either be a small drusen at an early stage or a
large drusen at a later stage. This point is not taken into account by a normal CNN applied on drusen segmentation. This can cause the network to miss
drusen that are particularly small or, conversely, drusen that exceed the network's receptive field (Fig. 3). In addition, retinal layers strictly follow the same
topological ordering and drusen has to appear strictly in-between OBRPE and
BM. In CNN models, contextual information and the spatial relation between
different anatomical parts of the retina might be overlooked by the small receptive field of a CNN. The limitations of receptive fields in a CNN is discussed in
more details in [8,9].

A solution is to increase the receptive field in the CNN architecture. This
could be approached in different ways, e.g. by a dilated convolution [10]). In
Pyramid scene parsing network (PSPNet), this is solved by a pyramid pooling
module [8]. Pyramid pooling is applying pooling with different window sizes.
The idea is instead of having one size pooling with common kernel size of $2 \times
2$ resulting in halved size feature maps, applying pyramid pooling layer with
different kernel size resulting in a sets of bins in a pyramid order (for example
1×1, 2×2, 3×3, 6×6). The coarsest pyramid level (1×1) resembles global
pooling that covers the entire image (see Fig. 1(e)). Spatial pyramid pooling is
also used in [9,11].

In [9,11,12] a spatial pyramid pooling layer is used once at the end of the
last convolutional layer of the network. In this paper we take one step further
and use a spatial pyramid pooling layer after each convolutional block of the
encoder of a standard U-Net. We also evaluate the result of segmenting three
classes instead of two classes, i.e. considering drusen as an additional, extra
class. Finally, we use a weighted loss function to train our proposed model. We

evaluate the performance of our approaches on the task of drusen and layer segmentation in retinas imaged with OCT. Results showed that the introduced model outperforms the baselines in term of Dice index of drusen segmentation, while also producing accurate delineations of the BM and the OBRPE surfaces.

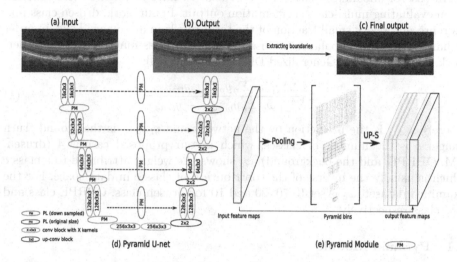

Fig. 1. (a) An input B-scan. (b) Corresponding network prediction output. (c) Final output after the post-processing. (d) The proposed model architecture. PM shows a pyramid module that is applied on the feature maps before they are passed to the next level. In the PM of each convolutional block on the decoder (pink PM), the reference size for output feature maps is half that of the input image size. In fact, the size of the feature maps in each layer of the decoder is halved. In the PM used for skip connections (Brown PM), the reference size is the same as the that of the input image. Therefor, the size of the feature maps passing through the skip connections does not change. (e) Five level pyramid module. The feature maps are gone through PM in order to have five-level bins. These five-level feature maps after up-sampling (UP-S) are concatenated with the original feature maps. (Color figure online)

2 Methods

U-Net [13] has proven to be a suitable architecture for medical images, as it uses skip-connections to pass the feature maps from the encoder at the same level during the reconstruction stage, which makes the model convenient for segmentation tasks where precise location is needed. Thus, we chose U-Net as a backbone for our proposed pyramid U-net with input image size of 256 × 256.

A retina OCT scan is comprised of sequential 2D B-scans. Usually, segmentation algorithms detect the drusen boundaries in B-scans by segmenting the outer RPE and BM surfaces, as opposed to segmenting the drusen directly. In order to provide more information to the network, in this work we define a four-class segmentation task: *Drusen, RPE region, BM region* and *Background* (Fig. 1(b)). This is our first implemented approach and we evaluate whether adding the extra class helps the network to learn how the drusen class interacts with the neighboring classes.

In case of unbalanced classes, it is crucial to have a weighted loss function when evaluating multi-class segmentation output. In our work, drusen class pixels represent a very small fraction of the total pixels in an image. Thus, in order to handle the class imbalance, we use the following loss function to train a network that is based on Generalized Dice Coefficient [14]:

$$-2\frac{\Sigma_{c=1}^{4}\ \omega_c\ \Sigma_n y_{pred_{cn}} y_{true_{cn}}}{\Sigma_{c=1}^{4}\ \omega_c\ \Sigma_n y_{pred_{cn}} + y_{true_{cn}}}.\tag{1}$$

where $y_{pred_{cn}}$ is the prediction by the network and $y_{true_{cn}}$ is the ground truth image. c is the number of classes, which in our proposed case is 4 (drusen, BM, OBRPE, and the background). ω_c shows the weight attributed to a class c which is usually the inverse of the contribution of class c in data space. For the examined dataset, ω_c is set to 70, 20 and 10 for drusen class, OBRPE class and BM class respectively.

2.1   Pyramid Module

Figure 1 shows the architecture of our proposed model. Each convolutional block is composed of two convolutional layers with 3×3 convolutions. Each convolutional block in the encoder is followed by one Pyramid Module (PM). A PM is composed of 5 different pooling levels with bins of size (1×1), (2×2), (3×3), (6×6) and (16×16). The five-level pyramid module forms five separate sets of feature maps, each with a different size. Thus, in the first level of the network there are 5 sets of $32 \times 256 \times 256$ feature maps (Fig. 1(e)), i.e., one series of feature maps for each pooling size. We apply average pooling with kernel size *pool_size* on these feature maps in each pyramid level in order to have results with bin size $(1 \times 1), (2 \times 2), (3 \times 3), (6 \times 6), (16 \times 16)$, respectively.

In each level, a series of feature maps is followed by a separate 1×1 convolutional to reduce the dimensionality of the feature maps to n. In this paper, n is set to $32/2 = 16$ for the bin $(2 \times 2$ bin) and to $32/4 = 8$ in the remaining pyramid levels $(1 \times 1, 3 \times 3, 6 \times 6$ and $16 \times 16)$. In each pyramid level, pooling kernel size will be calculated as: *pool_size* = $(input\_shape/bin\_size)$ in order to get feature maps with the target bin size. Since we are using U-Net as a baseline, where encoder uses 2×2 max pooling in each level of the network, we keep the feature maps at each level of the network the same size as those in the basic U-Net. Therefore, all the feature maps of the different bin sizes are combined with the feature maps obtained by the pooling with kernel size 2×2. The idea is that the feature maps from different bin sizes will add additional global context information to the main 2×2 pyramid. The same rule applies for the following levels. If N is the number of the feature maps in each level and n the desired number of the feature maps from a pooling bin $b \times b$, n is set to $N/2$ for pooling with size 2×2 and to $N/4$ for the rest of the pooling bins in the pyramid.

After applying the 1×1 convolution in each pooling level $p \times p$, there are 5 sets of feature maps of different sizes. In order to be able to concatenate these feature maps, each series of feature maps is up-sampled to the reference size of S. For the pyramid module in the decoder, S is set to $1/2 \times (input\_size)$. The feature maps at each level of the decoder are concatenated with the feature maps resulting from (2×2) max pooling (Fig. 2(a)). After concatenation, these pyramid feature maps are concurrently fed into the next layer. Conversely, for the pyramid modules on the skip connections S is set to $1 \times (input\_size)$. Therefore, these feature maps keep their original dimension (Fig. 2(b)). The output of a PM (original feature maps and feature maps from 5 level bins) are simultaneously passed through the skip connections to the matching layer in the decoder.

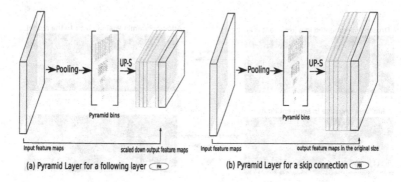

Fig. 2. (a) Pyramid module with the reference size of half of the input image size (b) Pyramid module with the reference size of the input image size.

The generalized Dice loss function was utilized for training the network. The predicted labels were regions for each target class (Fig. 1(b)). To acquire the final surfaces of the BM/OBRPE layers, a postprocessing strategy was applied. In each vertical column in the B-scan (called A-scan), the first row of activated pixels was extracted from the predicted BM region as the BM surface boundary. Similarly, in each vertical column in the B-scan (called A-scan), the last row of activated pixels was extracted from predicted OBRPE region as the OBRPE surface boundary.

3   Experimental Setup

Dataset. To train and evaluate the networks we use a private OCT dataset containing 425 OCT scans from 38 patients. We split the data into 34000 B-scans for training and validation (31 patients) and 7000 B-scans for testing (7 patients). Scans from the same subjects were always placed in the same set. Scans were acquired with Spectralis (Heidelberg Engineering, Heidelberg, Germany), which acquires anisotropic images with $1024 \times 97 \times 496$ voxels, each with the size of $5.7 \times 60.5 \times 3.87$ μm^3, and covering the field of view of $6 \times 6 \times 2$ mm^3.

Reference Standard. Each B-scan of every volume has been manually annotated in the following way. The Iowa Reference Algorithm [15] was first applied to generate a layer segmentation. The output was then manually corrected by an expert optometrist. Then, BM, OBRPE and the drusen regions are extracted from these annotations and used for training the network (Fig. 1).

Training Setup. Our method and the baselines [3,8] were trained with a batch size of 16 iterated for 50 times, using Adam optimization with an initial learning rate of $\eta = 10^{-5}$. Input B-scans are normalized to zero mean and unit variance and resized to 256 × 256 pixels. Based on Eq. 1, ω_c is set at 70, 20 and 10 respectively for drusen, *RPE region* and *BM region* in both baseline models and the introduced architecture.

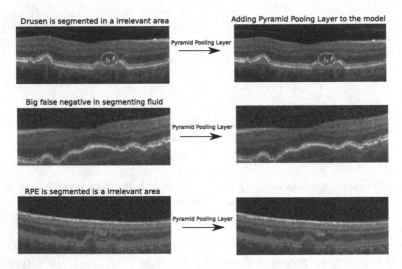

Fig. 3. (Left) Output of the basic U-Net [3] and (Right) output of the proposed model. First row: The basic U-Net has erroneously segmented the drusen as the bright areas above RPE, although there should be no drusen on top of the RPE. Second row (blue is false-negative drusen): in the basic U-Net's output, drusen region exceeds the network's receptive field. Third row (drusen in blue and RPE in red): The part of the image which is completely outside of the outer retina is segmented as RPE by the basic U-Net. (Color figure online)

4   Results

In order to evaluate our model, we compare it to several baselines. The first baseline is the standard U-Net architecture with two classes, BM and OBRPE, which has also been applied in the task of drusen segmentation by [3]. We denote this baseline as UNet-2C in Table 1. The second baseline is the U-Net with drusen introduced as an extra class. In this baseline, instead of extracting the area between BM and OBRPE as drusen, drusen is specifically segmented as an extra class. We denote this baseline as UNet-3C. Finally, our proposed model has in addition a spatial pyramid pooling layer at each level of the basic U-Net, and is denoted as UNet-PPM (Table 1).

Table 1. Quantitative evaluation. Patient-level mean Dice coefficient for drusen region segmentation and mean absolute error (MAE) in pixels for BM and OBRPE surface segmentation

Method	Dice (Drusen)	MAE (RPE)	MAE (BM)
UNet-2C [3]	70.25	1.42	1.35
UNet-3C	72.20	1.27	1.21
UNet-PPM	**74.73**	**0.79**	**0.71**

An example of segmentation output is shown in Fig. 3. It shows how the pyramid pooling method solves some fundamental issues in drusen segmentation by adding global contextual information to the feature maps which are being transferred through the network. We quantitatively evaluated the segmentation performance of the drusen, OBRPE and BM segmentation. Table 1 shows the results of this evaluation, per patient dice coefficient for drusen segmentation and mean absolute error for OBRPE and BM. In addition, Fig. 4 shows a box-plot of per patient dice coefficient for drusen and mean absolute error for BM and RPE segmentation. One can observe that by using the pyramid module, our proposed method was able to outperform the other baseline networks.

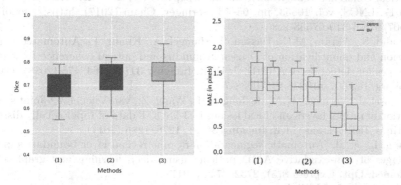

Fig. 4. Segmentation performance of different models on drusen (left), and OBRPE and BM (right). (1) UNet-2C [3]: U-Net with two classes, BM and OBRPE. (2) UNet-3C: original U-Net with three classes, BM, OBRPE and drusen. (3) UNet-PPL: the proposed model with three classes and pyramid pooling layers.

5   Discussion

Utilizing global spatial context is crucial for avoiding anatomically impossible segmentation such as finding drusen above RPE instead of below it. It is still a challenge to learn the plausible spatial relationships between object classes from a training dataset using statistical machine learning approaches. We proposed incorporating the pyramid pooling module into U-Net. The results showed that the proposed extension utilized the larger context for segmentation and clearly outperformed the baseline U-Net model. The proposed method is an important step towards the accurate quantification of drusen, crucial for the successful clinical management of patients with early AMD. Finally, given the widespread use of U-Net for medical image segmentation in general, the proposed extension would have an impact beyond its application in drusen segmentation.

Acknowledgment. This work was funded by the Christian Doppler Research Association, the Austrian Federal Ministry for Digital and Economic Affairs and the National Foundation for Research, Technology and Development. We thank the NVIDIA corporation for a GPU donation.

References

1. Wong, W.L., et al.: Global prevalence of age-related macular degeneration and disease burden projection for 2020 and 2040: a systematic review and meta-analysis. Lancet Glob. health **2**(2), e106–e116 (2014)
2. Schlanitz, F.G., et al.: Drusen volume development over time and its relevance to the course of age-related macular degeneration. Br. J. Ophthalmol. **101**(2), 198–203 (2017)
3. Gorgi Zadeh, S., et al.: CNNs enable accurate and fast segmentation of drusen in optical coherence tomography. In: Cardoso, M.J., et al. (eds.) DLMIA/ML-CDS -2017. LNCS, vol. 10553, pp. 65–73. Springer, Cham (2017). https://doi.org/10.1007/978-3-319-67558-9_8
4. Khalid, S., Akram, M.U., Hassan, T., Jameel, A., Khalil, T.: Automated segmentation and quantification of Drusen in Fundus and optical coherence tomography images for detection of ARMD. J. Digit. Imaging **31**(4), 464–476 (2018). https://doi.org/10.1007/s10278-017-0038-7
5. Novosel, J., Vermeer, K.A., de Jong, J.H., Wang, Z., van Vliet, L.J.: Joint segmentation of retinal layers and focal lesions in 3-D OCT data of topologically disrupted retinas. IEEE Trans. Med. Imaging **36**(6), 1276–1286 (2017)
6. Fang, L., et al.: Automatic segmentation of nine retinal layer boundaries in OCT images of non-exudative AMD patients using deep learning and graph search. Biomed. Opt. Express **8**(5), 2732–2744 (2017)
7. Shah, A., et al.: Multiple surface segmentation using convolution neural nets: application to retinal layer segmentation in OCT images. Biomedical Optics Express **9**(9), 4509–4526 (2018)
8. Zhao, H., Shi, J., Qi, X., Wang, X., Jia, J.: Pyramid scene parsing network. In: CVPR (2017)
9. He, K., Zhang, X., Ren, S., Sun, J.: Spatial pyramid pooling in deep convolutional networks for visual recognition. CoRR abs/1406.4729 (2014)

10. Yu, F., Koltun, V.: Multi-scale context aggregation by dilated convolutions. arXiv preprint arXiv:1511.07122 (2015)
11. Gu, Z., et al.: DeepDisc: optic disc segmentation based on atrous convolution and spatial pyramid pooling. In: Stoyanov, D., et al. (eds.) OMIA/COMPAY -2018. LNCS, vol. 11039, pp. 253–260. Springer, Cham (2018). https://doi.org/10.1007/978-3-030-00949-6_30
12. Zhao, R., et al.: Automated Drusen detection in dry age-related macular degeneration by multiple-depth, en face optical coherence tomography. Biomed. Opt. Express 8(11), 5049 (2017)
13. Ronneberger, O., Fischer, P., Brox, T.: U-Net: convolutional networks for biomedical image segmentation. In: Navab, N., Hornegger, J., Wells, W., Frangi, A. (eds.) MICCAI 2015. LNCS, vol. 9351, pp. 234–241. Springer, Cham (2015). https://doi.org/10.1007/978-3-319-24574-4_28
14. Crum, W.R., Camara, O., Hill, D.L.G.: Generalized overlap measures for evaluation and validation in medical image analysis. IEEE Trans. Med. Imaging 25(11), 1451–1461 (2006). Nov
15. Chen, X., Niemeijer, M., Zhang, L., Lee, K., Abràmoff, M.D., Sonka, M.: Three-dimensional segmentation of fluid-associated abnormalities in retinal OCT: probability constrained graph-search-graph-cut. IEEE-TMI 31(8), 1521–1531 (2012)

Deriving Visual Cues from Deep Learning to Achieve Subpixel Cell Segmentation in Adaptive Optics Retinal Images

Jianfei Liu, Christine Shen, Tao Liu, Nancy Aguilera, and Johnny Tam[✉]

National Eye Institute, National Institutes of Health, Bethesda, MD, USA
johnny@nih.gov

Abstract. Direct visualization of photoreceptor cells, specialized neurons in the eye that sense light, can be achieved using adaptive optics (AO) retinal imaging. Evaluating photoreceptor cell morphology in retinal diseases is important for monitoring the onset and progression of blindness, but segmentation of these cells is a critical first step. Most segmentation approaches focus on cell region extraction, without directly considering cell boundary localization. This makes it difficult to track cells that have ambiguous boundaries, which result from low image contrast, anisotropic cell regions, or densely-packed cells whose boundaries appear to touch each other. These are all characteristics of the AO images that we consider here. To address these challenges, we develop an AOSeg-Net method that uses a multi-channel U-Net to predict the spatial probabilities of the cell boundary and obtain cell centroid and region distribution information as a means for facilitating cell segmentation. Five-color theorem guarantees the separation of any touching cells. Finally, a region-based level set algorithm that combines all of these visual cues is used to achieve subpixel cell segmentation. Five-fold cross-validation on 428 high resolution retinal images from 23 human subjects showed that AOSegNet substantially outperformed the only other existing approach with Dice coefficients [%] of 84.7 and 78.4, respectively, and average symmetric contour distances [μm] of 0.59 and 0.80, respectively.

Keywords: U-Net · Level set segmentation · Adaptive optics · Five-color theorem · Cone photoreceptor neuron

1 Introduction

Adaptive optics (AO) retinal imaging can be used to directly visualize the morphology of photoreceptor cells in the living human eye [10]. Monitoring cell morphology can enhance the understanding of disease propagation at the cellular level. Subpixel cell segmentation is a prerequisite to monitor subtle cell changes as a one pixel error can cause up to 5% error in cell size measurements [5]. Accurate cell segmentation in AO images is hindered by low image contrast that

This is U.S. government work and not under copyright protection
in the U.S.; foreign copyright protection may apply 2019
H. Fu et al. (Eds.): OMIA 2019, LNCS 11855, pp. 86–94, 2019.
https://doi.org/10.1007/978-3-030-32956-3_11

often exists at cell boundaries (Fig. 1A). Anisotropic shading on opposite sides of cells requires special handling, and low pixel sampling necessitates subpixel segmentation in order to better monitor subtle changes in cell morphology. To date, there has only been one published method for automated cell segmentation in these AO images [10] (circularly-constrained active contour model, CCACM [5]). CCACM dynamically constructs circularly-shaped priors for each cell, which subsequently constrains active contours used in order to identify cell contours. Although CCACM achieved high accuracy when cells were loosely packed, it is prone to over-segmentation in the case of densely-packed regions where neighboring cells are very close together (cell crowding). This restricts the applicability of CCACM.

Recently, segmentation methods based on deep learning have shown substantial improvement over traditional image processing approaches [2,4,8,9,12,14]. The key challenge with AO images is separating crowded cells whose boundaries appear to touch each other. One approach would be to adaptively adjust weights at cell boundaries to train neural networks [8]. However, this approach is prone to over-segmentation of densely-packed cell regions. Contour-aware approaches [2,14] are effective ways to address cell crowding, but are prone to under-segmentation. To simplify the task of cell decrowding, joint cell segmentation and detection, combined with the use of star-convex polygons to represent cell shapes, was utilized [9]. Another possible solution is to post-process prediction results from the deep learning method by using conditional random fields [3]. However, this approach does not naturally achieve subpixel cell segmentation. Level set method [13] is an efficient means to address this issue by propagating active contours in a subpixel-level step. Five-color theorem has been combined with level sets to segment crowded cells [6]. However, identifying cell regions in terms of image intensity is unreliable, making the subsequent level set propagation in each colored region inefficient.

This paper introduces a combined approach incorporating deep learning and level sets for improving segmentation of photoreceptor cells in AO retinal images, particularly in dense regions where neighboring cell boundaries appear to touch each other. Our approach is called AOSegNet. It utilizes a multi-channel U-Net to simultaneously extract cell centroid, region, and contour visual cues, instead of only predicting cell regions as in the case of the traditional U-Net. Next, centroid and region cues from deep learning are combined to separate any clusters of touching cell regions into distinct, untouching regions, based on the five-color theorem. Finally, visual cues and colored regions are used to achieve subpixel cell segmentation. Five-fold cross-validation was performed to compare AOSegNet with existing methods. These results open up the possibility of monitoring subtle cellular changes that occur during neurodegenerative retinal diseases.

2   Methodology

AOSegNet consists of three components: a multi-channel U-Net (Fig. 1B) to extract cell cues (centroids, Fig. 1C; regions, Fig. 1D; and contours, Fig. 1E), a five-color theorem approach to separate touching cells (Fig. 1F), and a region-based level set method (Fig. 1G) to determine the final subpixel segmentation.

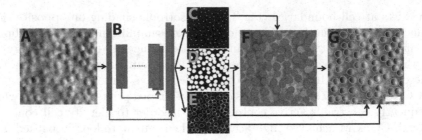

Fig. 1. Overview of AOSegNet for an example AO image of photoreceptors (A) using a multi-channel U-Net (B), which generates a set of visual cues consisting of centroids (C), regions (D), and contours (E). Cell centroid and region cues are used by the five-color theorem to separate touching cells. Red, blue and green regions represent distinct cells (F). A region-based level set segmentation is used to achieve subpixel cell segmentation (G). Scale bar, 20 μm (Color figure online)

2.1   Learning Cell Visual Cues

A multi-channel U-Net is leveraged to predict three cell visual cues: centroids, regions, and contours (Fig. 1B). Similar to the conventional U-Net [8], the multi-channel U-Net also consists of contracting and expanding paths (left and right sides, respectively). The contracting path is similar to a VGG network [11] that repeatedly applies 3×3 convolutions, followed by a rectified linear unit (ReLU) and 2×2 max pooling operations. The expanding path contains an upsampling series of the image feature map, which is a 2×2 convolution that concatenates image features from the contracting path, the upsampled feature map from the expanding path, and a ReLU.

Unlike the conventional U-Net [8] which only includes the region mask, the multi-channel U-Net contains centroid, region, and contour masks during training. A 3-channel label map is thus formulated, which improves prediction accuracy as they jointly constrain each other. It leads to a combinatorial loss function.

$$L = L_{centroid} + L_{region} + L_{contour} \qquad (1)$$

Visual cues of cell centroids and regions are represented as binary masks, $I(\mathbf{x}), \mathbf{x} \in \Omega$, where Ω is the image domain. The corresponding predictions are $\hat{I}(\mathbf{x})$. The loss function for each binary mask is formulated as the combination of binary cross entropy and Dice coefficient loss. In this way, $L_{centroid}$ and L_{region} can both be defined as

$$S(\mathbf{x}) = -\frac{1}{2} \sum_{i=1}^{2} I(\mathbf{x_i}) \log \hat{I}(\mathbf{x_i}) - \frac{2 \sum_{i=1}^{2} I(\mathbf{x_i})\hat{I}(\mathbf{x})}{\sum_{i=1}^{2} I^2(\mathbf{x_i}) + \sum_{i=1}^{2} \hat{I}^2(\mathbf{x_i})} \qquad (2)$$

To improve the accuracy of cell contour localization, the contour mask $I_c(\mathbf{x})$ is represented as a spatial density, which assigns probability values to image points near cell contours, assuming a Gaussian distribution, which results in

$$L_{contour} = S(\mathbf{x}) \exp\left(\left(I_c(\mathbf{x}) - \hat{I}_c(\mathbf{x}) \right)^2 / \sigma^2 \right) \tag{3}$$

where $\sigma = 0.5$ due to $I_c(\mathbf{x}) \in [0, 1]$. This term (Eq. 3) measures the intensity value changes between the labeled contour mask $I_c(\mathbf{x})$ and predicted contour mask $\hat{I}_c(\mathbf{x})$.

Altogether, the proposed multi-channel U-Net simultaneously predicts the probability masks of cell centroid, region, and contour cues (Figs. 1C-E) for a given AO retinal image (Fig. 1A).

2.2   Cell Decrowding

This step aims to extract cell regions that are clustered together in close proximity and separate them into groups of distinct regions. Although neighboring regions often touch, their simultaneously-learned centroids do not, which is key for efficient decrowding. Following Otsu's threshold method [7] to extract cell centroids and regions from their visual cue masks produced by the multi-channel U-Net, each centroid is used to identify its corresponding cell region through the watershed algorithm.

However, the extracted cell regions often contain segmentation errors due in large part to low pixel sampling and lack of subpixel accuracy. In order to achieve subpixel segmentation, cell regions within connected clusters of cells must first be disconnected from each another. We observe that cells within clusters can be separated based on the five-color theorem [1], which states that any 2D planar graph can be labeled with as few as five colors such that no neighbors have the same color. We can construct a planar graph with cell centroids as nodes, $V = \{v_1, v_2, \cdots, v_n\}$, with color $C(v)$ for each node. Cell centroids whose corresponding regions are connected to v_i are contained in $adj(v_i)$. The greedy coloring algorithm is used to assign a color to each cell region:

Algorithm 1. Greedy coloring

1: **for** $i = 1$ to n **do**:
2:     $c(v_i) := 0$
3: **end (for)**
4: **for** $i = 1$ to n **do**:
5:     Let $c(v_i)$ be the smallest $\mathbb{Z}^+$ s.t. $c(v_i) \notin \{c(v_j) : v_j \in adj(v_i)\}$
6: **end (for)**

We can thus separate connected cells into different groups with distinct, separated cells inside, as illustrated in Fig. 1F.

2.3   Guided Level Set Subpixel Segmentation

This step achieves subpixel segmentation by combining region and contour cues from the multi-channel U-Net with identity priors from the five-color theorem. Identity priors globally constrain level set propagation, and region and contour cues locally adjust level sets. A multiphase level set segmentation framework is defined as

$$E = \int_{\Omega} \sum_{i=1}^{m} \left(E_{region} + E_{contour} + E_{identity} \right) d\mathbf{x} \tag{4}$$

where $m \leq 5$ is constrained by the five-color theorem. Let $I_r(\mathbf{x})$ be the region cue mask from the multi-channel U-Net, and $\phi : \Omega \to \mathbb{R}$ be a signed distance function that represents the level set function.

$$E_{region} = (I_r - \mu_1)^2 H(\phi) + (I_r - \mu_2)^2 (1 - H(\phi)) \tag{5}$$

Here, $H(x)$ is the Heaviside function, with $H(x) = 1$ if $x \geq 0$; otherwise $H(x) = 0$. μ_1 and μ_2 are mean values of the mask regions inside and outside of the level set ϕ, respectively.

$$E_{contour} = c_1 F(I_c) |\nabla H(\phi)| \tag{6}$$

with $F(I_c) = 1 - I_c(\mathbf{x})$ because the contour cue mask is normalized to $I_c(\mathbf{x}) \in [0, 1]$ with large values at the cell boundary, where level set propagation should terminate.

Establishing signed distance functions ψ on different-colored cell regions (Fig. 1F) leads to the identity priors, which are defined as

$$E_{identity} = c_2 \left(H(\phi) - H(\psi) \right)^2 \tag{7}$$

Here, $c_1 = 2$ and $c_2 = 1.5$ represent scalar weights for balancing the level set framework for all of the images in this paper. Note that the identity priors only allow level sets to propagate near the image boundary that was predicted by the multi-channel U-Net, which reduces merging of cell regions contained within a certain multicolored region, while still achieving subpixel level cell segmentation.

The level set evolution equation is derived from Eq. 4 using Eqs. 5–7.

$$\frac{\partial \phi_i}{\partial t} = \delta(\phi_i) \left(-(I_r - \mu_i)^2 + \prod_{j \neq i} (1 - H(\phi_j)) \right)$$
$$+ c_1 \mathrm{div} \left(F(I_c) \frac{\nabla \phi_i}{|\nabla \phi_i|} \right) + 2c_2 (H(\phi_i) - H(\psi_i)) \right), 1 \leq i \leq m \tag{8}$$

Figure 1G shows the final cell segmentations computed using Eq. 8. Note that all touching clustered cells are successfully separated into individual cells with subpixel level accuracy.

2.4 Data Collection and Validation Methods

AO images of cone photoreceptors from 23 human subjects (age: 27.1 ± 8.8 years) were used to generate a total of 428 images (333×333 pixels), randomly selected from these subjects across different retinal regions. Note that AO images can vary substantially at different retinal regions of the same subject due to the variation of cone photoreceptor density, eye motion, and imaging conditions. Therefore, it is reasonable to have AO images from the same subject in both training and test datasets. Cones were manually labeled with subpixel accuracy by expert graders familiar with AO images for validation purposes. Five-fold cross-validation was performed to evaluate the accuracy and robustness of AOSegNet.

We compared segmentation results with CCACM [5], which is, to our knowledge, the only existing automated cell segmentation method for AO images of cone photoreceptors. Quantitative comparison was performed using six metrics: area overlap (AP), Dice coefficient (DC), area difference (AD), average symmetric contour distance (ASD), symmetric room mean square contour distance (RSD), and maximum symmetric absolute contour distance (MSD). Finally, cone diameters measured from our segmentation results were compared to previously published diameter measurements, including histological studies.

3 Experimental Results

3.1 Five-Fold Cross-Validation of Segmentation Accuracy and Robustness

Across each of the five folds, an average of 1343 corresponding cell regions were extracted by AOSegNet and CCACM. They were compared to each other and also to manually-labeled groundtruth. In all cases, AOSegNet performed substantially better than CCACM (Table 1). In each fold, the training time for the multi-channel U-Net was $\sim$6 h (2000 iterations; Microsoft Windows 7, Intel(R) core(TM) i7-6850K CPU, and dual NVIDIA GeForce GTX 1080 Ti GPUs). Following training, evaluation on each test dataset required less than 5 s per image.

Table 1. Segmentation accuracy comparison between AOSegNet and CCACM [5] over five-fold cross-validation

Method	AP (%)	DC (%)	AD (%)	ASD (μm)	RSD (μm)	MSD (μm)
AOSegNet	**74.2 ± 0.8**	**84.7 ± 0.6**	**19.9 ± 0.9**	**0.59 ± 0.02**	**0.70 ± 0.02**	**1.39 ± 0.04**
CCACM	66.0 ± 0.6	78.4 ± 0.4	26.9 ± 0.9	0.80 ± 0.02	0.98 ± 0.02	1.99 ± 0.05

Examples of segmentation results showed high cell segmentation accuracy on AO retinal images using AOSegNet (Fig. 2). Compared to CCACM, AOSegNet improved detection accuracy, and reduced both over- and under- segmentation (white arrows in Fig. 2). Our method combines spatial density contour cues as well as the five-color-theorem separation strategy, in order to accurately identify the contours of all photoreceptor cells.

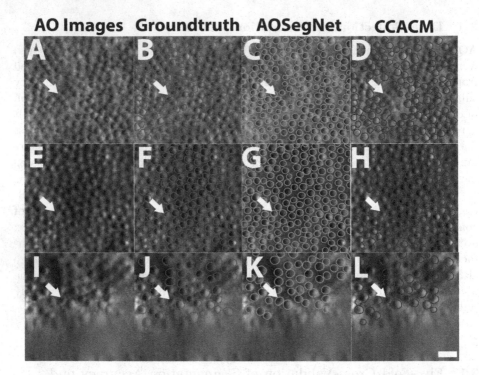

Fig. 2. Segmentation results on AO images varying in image quality and content. Compared to CCACM, AOSegNet improves detection accuracy (white arrows, top row), reduces over-segmentation (white arrows, center row), enhances under-segmentation (white arrows, bottom row), and performs well in the vicinity of image artifacts (lower portion, bottom row). Scale bar, 20 μm. (Color figure online)

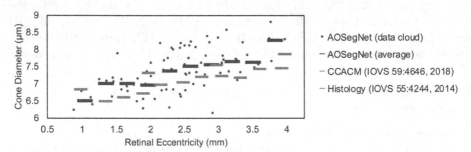

Fig. 3. Comparison of cone photoreceptor cell diameters generated with AOSegNet to those calculated with other methods. Each dot represents the average cone diameter measured within a single AO image (e.g. Fig. 1A). Since cone diameter varies depending on the location in the eye (retinal eccentricity), measured values were averaged every 0.3 mm in order to compare them to averaged values from previously-reported values. The average cone diameters measured were similar to previously-published values.

3.2   Cell Diameter Measurements

AOSegNet performed well across a test dataset consisting of 78 different AO images from healthy eyes. To demonstrate that the measurements were anatomically relevant, we computed cone photoreceptor cell diameters using the corresponding contours and compared them to those calculated with existing state-of-the-art methods (Fig. 3). Overall, AOSegNet measurements of cone diameters were similar to published values, including those measured from histological images. We also verified that the use of subpixel measurements improved accuracy: relative cell diameter differences of $7.9 \pm 0.3\%$ and $8.8 \pm 0.3\%$ were achieved for subpixel and pixel approaches, respectively, over the five folds, mean±SD.

4   Conclusion and Future Work

In this paper, we developed an AOSegNet for AO retinal images. A multi-channel U-Net was designed to simultaneously learn different types of visual cues (cell centroids, regions, and contours). These visual cues are used separately and in conjunction with each other in subsequent steps to intuitively improve segmentation performance. For example, cues integrated with the five-color theorem provide a simple solution to separate connected cell clusters, which substantially reduces segmentation errors when cells are crowded. By combining all learned and derived priors, we show that subpixel cell segmentation can be achieved. This subpixel representation is enabled in large part by the fact that cell contours were trained through a spatial density representation.

Five-fold cross-validation demonstrated that AOSegNet substantially outperforms the only existing AO photoreceptor cell segmentation method [5] across six different quantitative metrics (Table 1). These approaches will facilitate construction of normal databases of cell morphology in the living human eye, and will be useful for evaluating cell morphology in diseased eyes.

References

1. Appel, K., Haken, W.: Every planar map is four colorable. Illinois J. Math. **21**(3), 429–490 (1977)
2. Chen, H., Qi, X., Yu, L., et al.: DCAN: deep contour-aware networks for object instance segmentation from histology images. MedIA **36**, 135–146 (2017)
3. Chen, L., Papandreou, G., Kokkinos, I., et al.: DeepLab: semantic image segmentation with deep convolutional nets, atrous convolution, and fully connected CRFs. IEEE Trans. Pattern Anal. Mach. Intell. **40**(4), 834–848 (2018)
4. Gu, Z., Cheng, J., Fu, H., et al.: CE-Net: context encoder network for 2D medical image segmentation. IEEE Trans. Med. Imaging (2019, in press)
5. Liu, J., Jung, H., Dubra, A., Tam, J.: Cone photoreceptor cell segmentation and diameter measurement on adaptive optics images using circularly constrained active contour model. Invest. Ophthalmol. Vis. Sci. **59**(11), 4639–4652 (2018)

6. Nath, S.K., Palaniappan, K., Bunyak, F.: Cell segmentation using coupled level sets and graph-vertex coloring. In: Larsen, R., Nielsen, M., Sporring, J. (eds.) MICCAI 2006. LNCS, vol. 4190, pp. 101–108. Springer, Heidelberg (2006). https://doi.org/10.1007/11866565_13

7. Otsu, N.: A threshold selection method from gray-level histograms. IEEE Trans Cybern. **9**(1), 62–66 (1979)

8. Ronneberger, O., Fischer, P., Brox, T.: U-Net: convolutional networks for biomedical image segmentation. In: Navab, N., Hornegger, J., Wells, W.M., Frangi, A.F. (eds.) MICCAI 2015. LNCS, vol. 9351, pp. 234–241. Springer, Cham (2015). https://doi.org/10.1007/978-3-319-24574-4_28

9. Schmidt, U., Weigert, M., Broaddus, C., Myers, G.: Cell detection with star-convex polygons. In: Frangi, A.F., Schnabel, J.A., Davatzikos, C., Alberola-López, C., Fichtinger, G. (eds.) MICCAI 2018. LNCS, vol. 11071, pp. 265–273. Springer, Cham (2018). https://doi.org/10.1007/978-3-030-00934-2_30

10. Scoles, D., Sulai, Y., Langlo, C., et al.: In vivo imaging of human cone photoreceptor inner segments. Invest. Ophthalmol. Vis. Sci. **55**(7), 4244–4251 (2014)

11. Simonyan, K., Zisserman, A.: Very deep convolutional networks for large-scale image recognition. In: International Conference on Learning Representations (2015)

12. Valen, D.V., Kudo, T., Lane, K., et al.: Deep learning automates the quantitative analysis of individual cells in live-cell imaging experiments. PLoS Comput. Biol. **12**(11), e1005177 (2016). https://doi.org/10.1371/journal.pcbi.1005177

13. Vese, L., Chan, T.: A multiphase level set framework for image segmentation using the Mumford and Shah model. Int. J. Comput. Vis. **50**(7), 271–293 (2002)

14. Xu, Y., Li, Y., Wang, Y., et al.: Gland instance segmentation using deep multichannel neural networks. IEEE Trans. Biomed. Eng. **64**(12), 2901–2912 (2017)

Robust Optic Disc Localization by Large Scale Learning

Shilu Jiang[✉], Zhiyuan Chen[✉], Annan Li[✉], and Yunhong Wang[✉]

Beijing Advanced Innovation Center for Big Data and Brain Computing,
Beihang University, Beijing 100191, China
{shilu_jiang,dechen,liannan,yhwang}@buaa.edu.cn

Abstract. Since the optic disc (OD) is a main anatomical structure in retina, the localization of OD is an essential task in screening and diagnosing ophthalmic diseases. Many studies have been done for the automatic localization of OD but not reach a perfect performance yet. The bottleneck is lack of data and corresponding models that can handle with such big data. In this paper, we proposed an automatic OD localization method based on the hourglass network referenced from the human pose estimation task. Considering the lack of retina image databases, we also created a large retinal dataset of 85,605 images with manual OD bounding boxes. By learning from the large dataset, our deep network demonstrates excellent performance on OD localization. We also validated the proposed model on two public benchmarks, i.e. Messidor and ARIA datasets. Experiments show that it can achieve 100% accuracies on both datasets which clearly outperforms all the state-of-the-arts.

Keywords: Optic disc localization · Reinal image processing · Hourglass Network

1 Introduction

The optic disc (OD) or optic nerve head (ONH) is the point of exit for ganglion cell axons leaving the eye. Since it is the most salient feature of the eye fundus, it is actually used as a fiducial point in retinal image analysis. Therefore, OD localization is not only a prerequisite for disc relevant disease screening, such as glaucoma but also a necessary step for many other retinal image analysis tasks.

Conventional approaches either directly make use of the prior knowledge of OD shape or make estimation indirectly by the spatial context of vascular arch. Circle OD model has been used by Reza [15], Lu [9] and Abdullah et al. [1], while Yu et al. [19] and Kamble et al. [8] localize OD by the vessel convergence or profile. With the success of convolution neural network (CNN), classifier based detectors [2,16] are proposed to address the OD localization problem. However, unlike general object detection for multiple targets, there is only one optic disc in the eye fundus. Exhaustive search in such detector is unnecessary. Localization oriented encoder-decoder architecture is a more economic and suitable solution

© Springer Nature Switzerland AG 2019
H. Fu et al. (Eds.): OMIA 2019, LNCS 11855, pp. 95–103, 2019.
https://doi.org/10.1007/978-3-030-32956-3_12

for finding the OD. As a special kind of fully convolutional networks, U-Net [6] and its variant [3] have been investigated and show good performance.

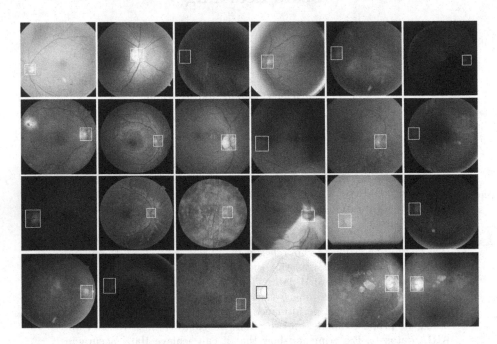

Fig. 1. Exemplar images and annotations. A large dataset for OD localization is built by marking the bounding box of Kaggle/EyePACS dataset.

However, the aforementioned studies are limited in three aspects: (1) simple handcrafted geometric models cannot deal with irregular variations of OD shape or the retinal vascular network; (2) although in theory learning based model can cover all the situations, existing models are trained on limited data, which makes them cannot deal with the low-quality images and rare cases shown in Fig. 1; (3) these models are relatively simple for the task of locating OD in various kinds of image captured in less-controlled scenarios.

To address these issues, a new OD localization model is presented in this paper. We annotate a large dataset with bounding box of the OD region and utilize it for training the deep neural networks. The proposed OD localization network is a symmetric encoder-decoder architecture which can learn features across all scales. It is also a recursive model with the residual learning modules. The OD location is estimated from the output probability heat map. Extensive experiments show the effectiveness and robustness of the proposed method.

The main contributions include: (1) A large retinal dataset of 85,605 images covering various challenging cases is constructed; (2) To facilitate the large dataset, an encoder-decoder network with deeper architecture and recursive mechanism is adopted to OD localization; (3) Extensive experiments show the superior performance over state of the arts.

2   Material

Although the importance of OD localization is well-recognized, relevant dataset is rare. Studies are carried out by using datasets designed for other retinal image analysis tasks, such as Messidor[1], ARIA[2] and STARE[3]. These datasets are limited in two aspects. Firstly, the image number is relative small. The largest image number of aforementioned dataset is 1,200, which is insufficient for training a reliable deep model. Secondly, the representativeness of data is limited. Since existing datasets used for OD localization are all disease oriented, the image quality has to be good for showing the details of lesion. However, as can be seen from Fig. 1, the real challenge of OD localization lies in low contrast, dimmed OD, confusing components, incomplete vascular arch etc. The number of such low quality image is very small in existing dataset, which limits both training result and the convincingness of evaluation. In other words, the good performance on such datasets does not necessarily mean good ability of coping with the aforementioned challenges.

Fortunately, the recent Kaggle/EyePACS dataset[4] provides a large number of raw fundus images. Al-Bander et al. [2] annotated the coordinates of OD center and fovea for 10,000 images. We argue that simple point annotation is inconsistent, since the shapes of OD are irregular and various. To build a reliable training set and benchmark, we annotated rectangle bounding box for all the images with complete visible OD in Kaggle/EyePACS dataset, which results in a large dataset consists of 85,605 eye fundus images. The bounding box provides not only an accurate and consistent annotation of OD center but also a more precise ground truth for evaluation. Exemplar image and annotation are shown in Fig. 1.

3   Method

The pipeline of proposed localization approach is illustrated in Fig. 2. At first, input fundus image is normalized to 256×256 by detecting the circle of field of view (FOV). Then the normalized image is fed to an encoder-decoder network with residual convolutional module, i.e., the hourglass net [13]. Output of the network is a probability map of OD location from which we can get the final coordinates of estimated OD center.

3.1   Preprocessing

Captured by different cameras, eye fundus images are different in size and aspect ratio. To facilitate the localization model, the input image needs to be normalized. Due to limited FOV, the camera can only focus on a small area on eye

---

[1] http://messidor.crihan.fr/index-en.php.
[2] http://www.eyecharity.com/aria-online.
[3] http://www.parl.clemson.edu/stare/nerve/.
[4] https://www.kaggle.com/c/diabetic-retinopathy-detection.

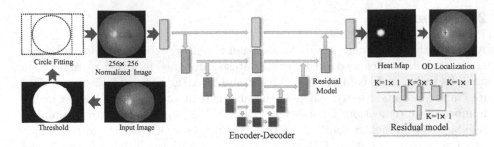

Fig. 2. Pipeline of proposed OD localization approach.

fundus, which results in a round shape visible area. We normalize the fundus image by fitting the circle and cropping the image according to the following criteria:

- All images are resized to an uniform height with preserved aspect ratio.
- The image is cropped to a square by trimming the margin columns according to the horizontal axis of the FOV circle.

In this work, we normalize the retinal image to 256×256. Examples of normalized images are shown in Fig. 1.

3.2   Optic Disc Localization by Hourglass Net

The hourglass net is originally proposed for human pose estimation [13], which is a similar task to OD localization. The backbone model of hourglass is residual module presented by He et al. [7]. The skip structure in the residual module makes the network deeper than existing models applied in OD localization. Consequently, a bigger and more flexible parameter space can be achieved, in which it is possible to fit an effective model to the large dataset introduced in Sect. 2.

The network input is a 256×256 retinal image, while the output is a probability heat map of OD location. Random mean and variance are used for data augmentation in normalization step. We use root mean square propagation (RMSprop) with a learning rate of $2.5e-4$ for optimization. At the training stage, the mark of OD center coordinate is transformed into a groundtruth heat map according to two dimensional Gaussian distributions. Then Mean-Squared Error (MSE) loss function is adopted to minimize the difference between output heat map and the ground truth.

$$MSE = \frac{1}{n} \sum_{i=1}^{n} (\widehat{x}_i - x_i)^2 \tag{1}$$

As for the final predicted location of OD center, we count the mean coordinate of all the pixels with the value greater than 70% of the maximum value in the output heat map.

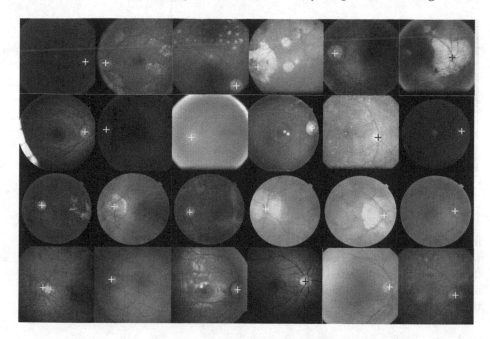

Fig. 3. Exemplar results of OD localization on EyePACS (first and second tow), MES-SIDOR (third row) and ARIA (bottom row).

4   Experiments

4.1   Settings

The proposed model is trained on the completely annotated Kaggle/EyePACS dataset described in Sect. 2. We strictly follow the train/test set division of the diabetic retinopathy detection competition, which results in a training set of 33,799 images and a test set of 51,806 images. The proposed method is also evaluated on Messidor and ARIA dataset for comparison with the state of the art. Messidor consists of 1,200 images with various resolution of 1440×960, 2240×1488 and 2304×1536, while ARIA includes 120 images of an uniform size 768×576. OD in both datasets are labeled by region mask. The implementation is based on PyTorch [14] and a machine with two NVIDIA GeForce GTX TITAN X graphic cards. The batch size is set to 8.

The accuracy of OD localization is measured by two kinds of methods. The first is morphologic region: the predicted OD location is considered successful if it falls in the OD region. Depends on the annotation, the OD region can be rectangle bounding box, oval area or the manual fine contour/mask. Besides that radius-dependent accuracy is also adopted, in which $\frac{1}{8}$, $\frac{1}{4}$, $\frac{1}{2}$ and one radius are investigated. On EyePACS the radius is determined by half of the long edge of rectangle, while on Messidor we adopt the settings of prior arts, in which the radius is set to 68, 103 and 109 pixels for 1440×960, 2240×1488 and 2304×1536

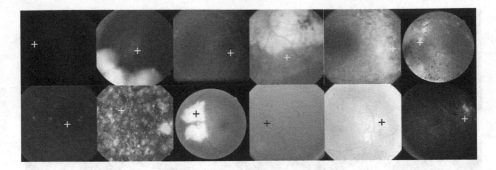

Fig. 4. Failure cases on EyePACS.

image respectively. On ARIA the radius is calculated by half of the OD mask width. For calculating the error distance, the OD center is also needed in all dataset which is calculated by the geometric center of OD mask respectively. As a measure of computational efficiency, we also calculated the average processing time per image.

4.2   Results

Exemplar results of OD localization on three datasets are shown in Fig. 3, and the quantitative comparisons are presented in Table 1 and 2 respectively. As can be seen, the presented new test set (EyePACS) is much bigger than existing datasets, and includes many challenging images. Even though, the proposed method can still achieve high accuracies. The effectiveness of large-scale learning is well demonstrated. It is impressive that the localization results are much more precise than Al-Bander et al. [2], which is also trained and tested on EyePACS. When the acceptable error distance to OD center is limited to $\frac{1}{4}$ radius, the accuracy of [2] degrades to 51.9%, while that of our method can be still above 99%. In our work, the average *Euclidean Distance (ED)* to the true OD center can be as small as 1.76 pixel on the 256×256 image.[5] This phenomenon implies that the proposed model is more stable.

To compare with the existing approaches, we also perform experiments on Messidor and ARIA. As can be seen, our model achieves perfect results and consistently outperforms the state of the art in all metrics. Due to the difference of camera, adapting model to another dataset usually leads to performance decline. However, our model trained on EyePACS shows great generalization performance on both Messidor and ARIA. Based on morphologic region metric, the accuracy maintains 100% on both datasets in all kinds of OD regions. When based on radius-dependent measurement even the acceptable error distance to OD center is limited to $\frac{1}{4}$ radius, the accuracy of our method is still above 99%. Besides the accuracy we also evaluate the computational efficiency, the average processing time per image is 5 milliseconds which shows that the proposed

---

[5] Distances to OD center on Messidor and ARIA are calculated in original resolution.

Table 1. Performance comparison based on morphologic regions (%).

Dataset	Method	Box	Oval	Contour	ED	Time (ms)
Messidor(1,200)	Sadhukhan et al. [16]	–	–	98.75	14	**0.15**
Messidor(1,200)	Lu [10]	–	–	99.75	–	5000
Messidor(600)	Gu et al. [6]	–	–	99.83	–	–
Messidor(1,200)	Yu et al. [19]	–	–	99.67	–	–
Messidor(1,200)	Abdullah et al. [1]	–	–	99.25	–	–
Messidor(1,200)	R. Kamble et al. [8]	–	–	99.75	-	520
Messidor(1,200)	Proposed	**100**	**100**	**100**	**8.03**	5
ARIA(120)	Lu [10]	–	–	97.5	6	5000
ARIA(120)	Proposed	**100**	**100**	**100**	**4.14**	5
EyePACS(51,806)	Proposed	99.8	99.88	–	1.76	5

Table 2. Performance comparison of OD radius-dependent accuracies (%).

Dataset	Method	$\frac{1}{8}$R	$\frac{1}{4}$R	$\frac{1}{2}$R	R	Time (ms)
Messidor(1,200)	Al-Bander et al. [2]	–	83.6	95	97	7
Messidor(1,200)	Giachetti et al. [5]	–	–	–	99.66	5000*
Messidor(1,200)	Yu et al. [18]	–	–	–	99	4700
Messidor(1,200)	Yu et al. [17]	–	–	99.08	98.24	–
Messidor(1,200)	Marin et al. [11]	87.33	97.75	99.50	99.75	–
Messidor(1,136)	Meyer et al. [12]	65.58	93.57	97.10	98.94	–
Messidor(1,200)	Gegundez-Arias et al. [4]	80.42	93.92	96.08	96.92	940
Messidor(1,200)	Proposed	81.58	**99.58**	**100**	**100**	5
ARIA(120)	Proposed	60.00	97.50	100	100	5
EyePACS(3,000)	Al-Bander et al. [2]	–	51.9	87.4	96.7	7
EyePACS(51,806)	Proposed	84.47	**99.42**	**99.82**	**99.89**	5

*Time of Giachetti et al. includes both OD and fovea detection.

method can satisfy the requirement of real-time processing. It should be pointed out that the speeds of compared approaches shown in Table 1 and 2 are directly cited from corresponding articles, in which the hardwares are different.

In order to clarify the importance of our large dataset, we also conducted an comparative experiment of different fractions of dataset used for training. The results in Table 3 shows that the larger dataset will leads to better performance, which proves the significance of our large dataset.

Some failure cases on EyePACS dataset are shown in Fig. 4. As can be seen, they are either low-quality images or the rare cases. Even for human, identifying the OD location is difficult. We argue that such images can be easily classified for special treatment since their holistic appearances are very different from the

Table 3. Performance comparison of different training sets.

Fraction	EyePACS(51806)	Messidor(1200)	ARIA(120)
1% (865)	97.4	98.92	97.50
5% (5000)	99.0	99.17	99.17
20% (16896)	99.5	99.83	99.17
40% (33789)	99.8	100	100

*The performance on EyePACS is based on box region and on others is based on contour region.

normal ones. In other words, in a practical system, such low-quality or special images are usually excluded by the quality assessment.

5 Conclusion

In this paper, we propose an approach for robust optic disc localization in color fundus images. To achieve a large and representative dataset, 85,605 fundus images are annotated with OD bounding box. To facilitate this large dataset, an encoder-decoder network with deep residual structure and recursive learning mechanism is adopted for robust OD localization. Experimental results show that the learned model is efficient and can clearly outperform state of the arts.

Acknowledgment. This work was supported by the Foundation for Innovative Research Groups through the National Natural Science Foundation of China under Grant 61421003.

References

1. Abdullah, M., et al.: Localization and segmentation of optic disc in retinal images using circular Hough transform and grow-cut algorithm. PeerJ **4**(3), e2003 (2016)
2. Al-Bander, B., et al.: Multiscale sequential convolutional neural networks for simultaneous detection of fovea and optic disc. BSPC **40**, 91–101 (2018)
3. Fu, H., et al.: Joint optic disc and cup segmentation based on multi-label deep network and polar transformation. TMI **37**(7), 1597–1605 (2018)
4. Gegundez-Arias, M.E., et al.: Locating the fovea center position in digital fundus images using thresholding and feature extraction techniques. CMIG **37**(5–6), 386–393 (2013)
5. Giachetti, A., et al.: The use of radial symmetry to localize retinal landmarks. CMIG **37**(5–6), 369–376 (2013)
6. Gu, Z., et al.: Automatic localization of optic disc using modified U-Net. In: ICCCV, pp. 79–83 (2018)
7. He, K., et al.: Deep residual learning for image recognition. In: CVPR, pp. 770–778 (2016)
8. Kamble, R., et al.: Localization of optic disc and fovea in retinal images using intensity based line scanning analysis. CBM **87**, 382–396 (2017)

9. Lu, S., et al.: Accurate and efficient optic disc detection and segmentation by a circular transformation. TMI **30**(12), 2126–2133 (2011)
10. Lu, S., et al.: Automatic optic disc detection from retinal images by a line operator. TBME **58**(1), 88–94 (2011)
11. Marin, D., et al.: Obtaining optic disc center and pixel region by automatic thresholding methods on morphologically processed fundus images. CMPB **118**(2), 173–185 (2015)
12. Meyer, M.I., Galdran, A., Mendonça, A.M., Campilho, A.: A pixel-wise distance regression approach for joint retinal optical disc and fovea detection. In: Frangi, A.F., Schnabel, J.A., Davatzikos, C., Alberola-López, C., Fichtinger, G. (eds.) MICCAI 2018. LNCS, vol. 11071, pp. 39–47. Springer, Cham (2018). https://doi.org/10.1007/978-3-030-00934-2_5
13. Newell, A., Yang, K., Deng, J.: Stacked hourglass networks for human pose estimation. In: Leibe, B., Matas, J., Sebe, N., Welling, M. (eds.) ECCV 2016. LNCS, vol. 9912, pp. 483–499. Springer, Cham (2016). https://doi.org/10.1007/978-3-319-46484-8_29
14. Paszke, A., et al.: Automatic differentiation in PyTorch. In: NIPS-W (2017)
15. Reza, M.N., et al.: Automatic detection of optic disc in color fundus retinal images using circle operator. BSPC **45**, 274–283 (2018)
16. Sadhukhan, S., et al.: Optic disc localization in retinal fundus images using faster R-CNN. In: EAIT, pp. 1–4 (2018)
17. Yu, H., et al.: Fast localization of optic disc and fovea in retinal images for eye disease screening. In: CAD, vol. 7963, p. 796317 (2011)
18. Yu, H., et al.: Fast localization and segmentation of optic disk in retinal images using directional matched filtering and level sets. ITB **16**(4), 644–657 (2012)
19. Yu, T., et al.: Automatic localization and segmentation of optic disc in fundus image using morphology and level set. In: ISMICT, pp. 195–199 (2015)

The Channel Attention Based Context Encoder Network for Inner Limiting Membrane Detection

Hao Qiu[1,2], Zaiwang Gu[3], Lei Mou[2], Xiaoqian Mao[2], Liyang Fang[2], Yitian Zhao[2], Jiang Liu[3], and Jun Cheng[2(✉)]

[1] School of Mechatronic Engineering and Automation,
Shanghai University, Shanghai, China
[2] Cixi Institute of Biomedical Engineering,
Ningbo Institute of Industrial Technology,
Chinese Academy of Sciences, Ningbo, China
Chengjun@nimte.ac.cn
[3] Department of Computer Science and Engineering,
Southern University of Science and Technology, Shenzhen, China

Abstract. The optic disc segmentation is an important step for retinal image based disease diagnosis such as glaucoma. The inner limiting membrane (ILM) is the first boundary in the OCT, which can help to extract the retinal pigment epithelium (RPE) through gradient edge information to locate the boundary of the optic disc. Thus, the ILM layer segmentation is of great importance for optic disc localization. In this paper, we build a new optic disc centered dataset from 20 volunteers and manually annotated the ILM boundary in each OCT scan as ground-truth. We also propose a channel attention based context encoder network modified from the CE-Net [1] to segment the optic disc. It mainly contains three phases: the encoder module, the channel attention based context encoder module, and the decoder module. Finally, we demonstrate that our proposed method achieves state-of-the-art disc segmentation performance on our dataset mentioned above.

Keywords: Disc segmentation · ILM layer detection · Channel attention based context encoder

1 Introduction

Glaucoma is the second leading cause of blindness globally, which may result in vision loss and irreversible blindness. The number of people suffering from glaucoma is estimated to increase to 80 million in 2020 [2]. As the disease progresses asymptomatic in the early stages, the majority of the patients are unaware until an irreversible visual loss occurs. Thus, early diagnosis and treatment for glaucoma is utmost essential for preventing the deterioration of vision. While there

H. Fu et al. (Eds.): OMIA 2019, LNCS 11855, pp. 104–111, 2019.
https://doi.org/10.1007/978-3-030-32956-3_13

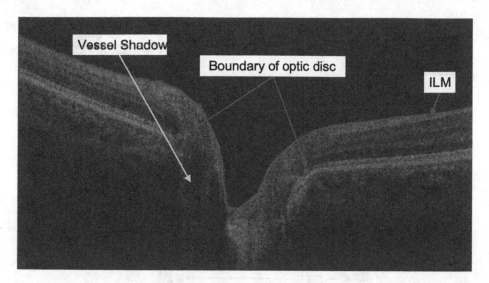

Fig. 1. Optic nerve head structure in a cropped OCT slice. The red curve denotes the ILM boundary. The blue points refer to the boundary points of the optic disc. ILM: Inner limiting membrane. (Color figure online)

are various approaches to diagnose glaucoma such as vessel distribution, FFT/B-spline coefficients, most of the known literature has endeavoured to assess the cup-to-disc ratio (CDR).

There have been a number of attempts at automatically detecting the optic disc in ocular images. Many proposed optic disc detection approaches concentrate on segmenting the optic region in color fundus images. For example, Liu *et al.* [3] proposed Variational level set approach for segmentation of optic disc without reinitialization. Xu *et al.* [4] employed the deformable model technique through minimization of an energy function to detect the disc. Cheng *et al.* [5] used the state-of-the-art self-assessed disc segmentation method combined three methods to segment the disc. However, these proposed approaches face challenges when the optic disc does not have a distinct color in the fundus image.

Optical coherence tomography (OCT), an important retinal imaging method with non-invasive, high-resolution characteristics, provides the fine structure within the human retina [6]. A single image of OCT slice is shown in Fig. 1. Some optic disc segmentation methods are applied to 3-D OCT volumes. For example, Lee *et al.* [7] applied a K-NN classifier to segment the optic disc cup and neuroretinal. Fu *et al.* [8] provided a Low-rank reconstruction to automatically detect optic disc in OCT slices.

With the development of convolutional neural network (CNN) in image and video processing [9], automatic feature learning algorithms using deep learning have emerged as feasible approaches and are applied to handle the image analysis. Recently, some deep learning based segmentation algorithms have been proposed to segment medical images [10], [1]. Based on the U-Net, a recent

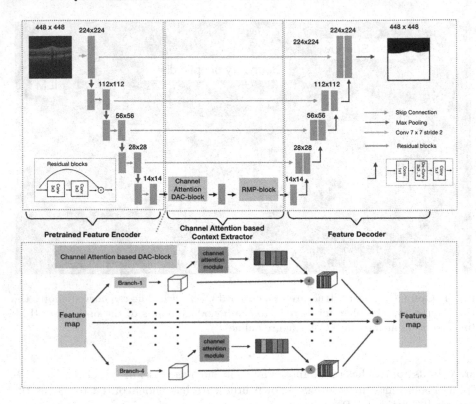

Fig. 2. Illustration of the proposed CACE-Net. Firstly, the images are fed into a feature encoder module, where the residual network (ResNet) block was employed as the backbone for each block, and then followed by a max-pooling layer to increase the receptive field for better extraction of global features. Then the features from the encoder module are fed into the proposed channel attention based context encoder module. Finally, the decoder module was used to enlarge the feature size and output a mask, the same size as the original input.

popular medical image segmentation architecture, CE-Net employs multi-scale atrous convolution and pooling operations to improve the segmentation performance. And it achieves some state-of-the-art performance in some medical image segmentation tasks, such as optic disc segmentation and OCT layers segmentation. The original context extractor module in CE-Net was consist of a dense atrous convolution (DAC) module and a residual multi-kernel pooling (RMP) module. However, the original DAC and RMP accounted for abundant channels to enrich the semantic features representations. Each channel of the features at the classification layer can be regarded as a specific-class response since we add the supervision signal on this layer. These abundant channels could be further embedded to produce the global distribution of channel-wise feature responses. In this paper, in order to extract more high-level semantic features, we introduce the channel attention mechanism to enhance the context extractor module of the

CE-Net, and propose a channel attention based context encoder network (called CACE-Net) for inner limiting membrane detection.

The major contributions of this work are summarized as follows:

(1) We annotate 20 3D-OCT scans (both of them are right eye scans) centered at optic disc.
(2) we leverage the ability of CACE-Net to accurately segment the inner limiting membrane (ILM) in our dataset, which is defined as the boundary between the retina and the vitreous body. This is necessary for our further work to detect the optic disc boundary points. The segmentations on database of OCT images are demonstrated to be superior to those from some known state-of-the-art methods. And we will release our code and dataset on Github later.

2   Proposed Method

The CE-Net [1] achieves the state-of-the-art performances in some 2D medical image segmentation tasks, such as optic disc segmentation, retinal vessel detection, lung segmentation and cell contour extraction. The proposed CACE-Net is modified from the CE-Net, which mainly contains three phases: the encoder module, the channel attention based context encoder module, and the decoder module, as shown in Fig. 2. The feature encoder module includes four encoder blocks, and the residual network (ResNet) block was employed as the backbone for each block, and then followed by a max-pooling layer to increase the receptive field for better extraction of global features. Then the features from the encoder module are fed into the proposed channel attention based context encoder module. Finally, the decoder module was used to enlarge the feature size and output a mask, the same size as the original input.

2.1   Channel Attention Based Context Extractor Module

The original context extractor module in CE-Net [1] employed four cascade branches with multi-scale atrous convolution to capture multi-scale semantic features, followed by various size pooling operations to further encode the multi-scale context features. This module accounts for abundant channels to enrich the semantic features representations, which could be further embedded to generate the global distribution of channel-wise feature responses. Therefore, motivated by the SE-Net [11], we propose a channel attention based context extractor module, introducing the relationship between channels.

In this section, we mainly introduce how to exploit the interdependencies of channel maps, as illustrated in Fig. 2. The proposed channel attention based context extractor module employs channel attention mechanism to allow the network to perform feature recalibration of aggregated context features, with the basis of original DAC block. Specially, the CACE module utilizes four cascade branches with multi-scale atrous convolution and channel attention module, to gain high-level features.

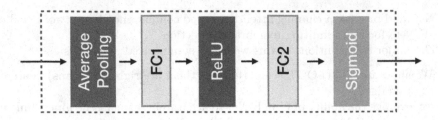

Fig. 3. Illustration of the channel attention module.

As illustrated in Fig. 3, the extracted feature map $F \in \mathbb{R}^{C \times H \times W}$ in channel attention module is first calculated directly by the global average pooling to generate channel-wise statistics $z \in \mathbb{R}^C$:

$$z_c = \frac{1}{H \times W} \Sigma_{i=1}^{H} \Sigma_{j=1}^{W} f_c(i, j) \tag{1}$$

where $H \times W$ represents the spatial dimensions of features and C is the number of channels. Then, the two linear transformations W_1, W_2 and a sigmoid activation function σ are employed to obtain the squeeze and excitation statistics $s \in \mathbb{R}^C$:

$$s_c = \sigma(W_2 \delta(W_1 z_c)) \tag{2}$$

where δ refers to the ReLU function, $W_1 \in \mathbb{R}^{\frac{C}{r} \times C}$ and $W_2 \in \mathbb{R}^{C \times \frac{C}{r}}$. Finally, a matrix multiplication between the statistics $s \in \mathbb{R}^C$ and the feature $F \in \mathbb{R}^{C \times H \times W}$ is added to obtain the final output in each branch of the proposed channel attention DAC module, followed by the RMP block for further context information with multi-scale pooling operations.

2.2   Feature Decoder

Instead of directly upsampling the features to the original image dimensions, we follow the CE-Net [1] to introduce a feature decoder module that restores the dimensions of the high level semantic features layer by layer. In each layer, we use ResNet block as the backbone of the decoder block which is followed by a 1 × 1 convolution, a 3 × 3 transposed convolution, a 1 × 1 convolution. Similar to U-Net [12], we add a skip connection between each layer of the encoder and decoder. Finally, the feature decoder module could generate the prediction of the same size as the original input.

2.3   Boundary Extractor

The main goal of this method is to detect internal limiting membrane. Therefore, we need to turn the segmentation prediction to a boundary line, which corresponds to the internal limiting membrane. We remove the small connected components to denoise the segmentation prediction, adopting the morphology method. After this post processing operation, we achieve the final boundary corresponding to the internal limiting membrane between the retina and the vitreous body.

2.4   Loss Function

In this method, we choose binary cross-entropy loss as our loss function $\mathcal{L}_B$, since the method just needs to predict the binary outputs. The binary cross-entropy loss is as follows:

$$\mathcal{L}_B = -\mathbb{E}_{x \sim p_{data}}[y \cdot \log(D(x)) + (1 - y) \cdot \log(1 - D(x))], \tag{3}$$

where y represents the ground truth, and $D(x)$ is the prediction.

3   Experiment Results

3.1   Dataset and Metric

20 3D-OCT scans (both of them are right eye scans) centered at optic disc were collected from 20 volunteers. Each OCT scan consisted of 885×512 image resolution. While there exist methods for extracting multiple retinal layers from OCT slices, only ILM layer boundaries is needed in our paper. The ILM is defined as the boundary between the retina and the vitreous body, which is the first boundary of retinal OCT. The ground-truth optic disc boundary of a 3D-OCT volume is obtained by first manually labeling the optic disc points in each optic disc centered slice (with a trained labeler and two experts for quality control). These labeled points were then to generate the ground-truth optic disc boundary. In our paper, we also randomly take 10 people's images for training, and others for testing. In this paper, we follow the same partition of the data set to train and test our models.

Following the previous approaches [1], we compute the mean absolute error (mae) between prediction and ground truth as the metric to evaluate the accuracy of segmentation algorithms.

$$error = \frac{1}{n} \sum_{i=1}^{n} |y_i - Y_i| \tag{4}$$

where y_i represents the i_{th} pixel predicted value of one surface, and Y_i represents that of ground truth.

3.2   Implementation Details

The proposed CACE-Net was implemented on PyTorch library with the NVIDIA GPU. We choose stochastic gradient descent (SGD) optimization, other than adaptive moment estimation (Adam) optimization. We use SGD optimization since recent studies [13] show that SGD often achieves a better performance, though the Adam optimization convergences faster. The initial learning rate is set to 0.001 and a weight decay of 0.0001. We use poly learning rate policy where the learning rate is multiplied by $\left(1 - \frac{iter}{max\_iter}\right)^{power}$ with power 0.9. All training images are rescaled to 448×448.

In order to demonstrate conclusively the superiority of the proposed method over the other methods, we compare our method with two algorithms for the ILM segmentation:

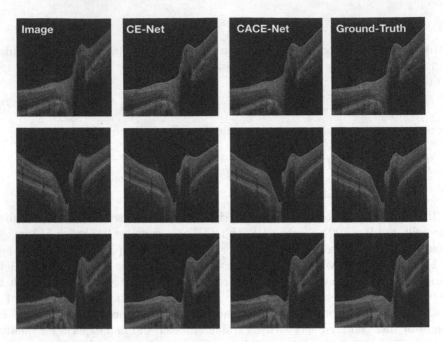

Fig. 4. Sample results of the ILM segmentation. From left to right: original images, CE-Net, CACE-Net and Ground-Truth

(1) U-net, a popular neural network architecture for biomedical image segmentation tasks.
(2) CE-Net [1], which achieves the state-of-the-art performances in some 2D medical image segmentation tasks, such as optic disc segmentation, retinal vessel detection, lung segmentation and cell contour extraction.

3.3 Results and Discussion

As can be seen in Table 1, we show the performances of three optic disc segmentation algorithms. Compared with other state-of-the-art optic disc segmentation methods, our CACE-Net outperforms the other algorithms based on deep learning image processing method. From the comparison shown in Table 1, the CACE-Net achieves 2.199 in the mean absolute error, better than the U-Net. From the comparison between CE-Net [1] and our CACE-Net, we also observe that there is a drop of the mean absolute error by 10.8% from 2.467 to 2.199.

Table 1. Performance comparison of the ILM detection (mean $\pm$ standard deviation)

Method	U-Net	CE-Net	CACE-Net
error	6.404 ± 16.407	2.467 ± 1.989	2.199 ± 1.471

We also show three sample results in Fig. 4 to visually compare our method with the most competitive methods, CE-Net. The comparison images show that our method obtain more accurate segmentation results.

4 Conclusion

In this paper, we have built a manually labeled OCT dataset and proposed an effective architecture for segmenting the ILM layer in our OCT dataset. The proposed CACE-Net achieves the mean absolute error of 2.199 in our dataset, better than other methods.

References

1. Gu, Z., et al.: CE-NET: context encoder network for 2D medical image segmentation. IEEE Trans. Med. Imaging (2019)
2. Tham, Y.C., Li, X., Wong, T.Y., Quigley, H.A., Aung, T., Cheng, C.Y.: Global prevalence of glaucoma and projections of glaucoma burden through 2040: a systematic review and meta-analysis. Ophthalmology **121**, 2081–2090 (2014)
3. Liu, J., et al.: Automatic glaucoma diagnosis through medical imaging informatics. J. Am. Med. Inform. Assoc. **20**, 1021–1027 (2013)
4. Xu, J., Chutatape, O., Sung, E., Zheng, C., Kuan, P.C.T.: Optic disk feature extraction via modified deformable model technique for glaucoma analysis. Pattern Recogn. **40**, 2063–2076 (2007)
5. Cheng, J., Yin, F., Wong, D.W.K., Tao, D., Liu, J.: Sparse dissimilarity-constrained coding for glaucoma screening. IEEE Trans. Biomed. Eng. **62**, 1395–1403 (2015)
6. Schmitt, J.M.: Optical coherence tomography (OCT): a review. IEEE J. Sel. Top. Quantum Electron. **5**, 1205–1215 (1999)
7. Lee, C.S., Tyring, A.J., Deruyter, N.P., Wu, Y., Rokem, A., Lee, A.Y.: Deep-learning based, automated segmentation of macular edema in optical coherence tomography. Biomed. Opt. Express **8**, 3440–3448 (2017)
8. Fu, H., Xu, D., Lin, S., Wong, D.W.K., Liu, J.: Automatic optic disc detection in OCT slices via low-rank reconstruction. IEEE Trans. Biomed. Eng. **62**, 1151–1158 (2014)
9. Krizhevsky, A., Sutskever, I., Hinton, G.E.: ImageNet classification with deep convolutional neural networks. In: Advances in Neural Information Processing Systems, pp. 1097–1105 (2012)
10. Gu, Z., et al.: DeepDisc: optic disc segmentation based on atrous convolution and spatial pyramid pooling. In: Stoyanov, D., et al. (eds.) OMIA/COMPAY -2018. LNCS, vol. 11039, pp. 253–260. Springer, Cham (2018). https://doi.org/10.1007/978-3-030-00949-6_30
11. Hu, J., Shen, L., Sun, G.: Squeeze-and-excitation networks. In: Proceedings of the IEEE Conference on Computer Vision and Pattern Recognition, pp. 7132–7141 (2018)
12. Ronneberger, O., Fischer, P., Brox, T.: U-Net: convolutional networks for biomedical image segmentation. In: Navab, N., Hornegger, J., Wells, W.M., Frangi, A.F. (eds.) MICCAI 2015. LNCS, vol. 9351, pp. 234–241. Springer, Cham (2015). https://doi.org/10.1007/978-3-319-24574-4_28
13. Keskar, N.S., Socher, R.: Improving generalization performance by switching from Adam to SGD. arXiv preprint arXiv:1712.07628 (2017)

Fundus Image Based Retinal Vessel Segmentation Utilizing a Fast and Accurate Fully Convolutional Network

Junyan Lyu, Pujin Cheng, and Xiaoying Tang$^{(\boxtimes)}$

Southern University of Science and Technology, Shenzhen, China
tangxy@sustech.edu.cn

Abstract. Monitoring the condition of retinal vascular network based on a fundus image plays a vital role in the diagnosis of certain ophthalmologic and cardiovascular diseases, for which a prerequisite is to segment out the retinal vessels. The relatively low contrast of retinal vessels and the presence of various types of lesions such as hemorrhages and exudate nevertheless make this task challenging. In this paper, we proposed and validated a novel retinal vessel segmentation method utilizing *Separable Spatial and Channel Flow* and *Densely Adjacent Vessel Prediction* to capture maximum spatial correlations between vessels. Image pre-processing was conducted to enhance the retinal vessel contrast. Geometric transformations and overlapped patches were used at both training and prediction stages to effectively utilize the information learned at the training stage and refine the segmentation. Publicly available datasets including DRIVE and CHASE_DB1 were used to evaluate the proposed approach both quantitatively and qualitatively. The proposed method was found to exhibit superior performance, with the average areas under the ROC curve being 0.9826 and 0.9865 and the average accuracies being 0.9579 and 0.9664 for the aforementioned two datasets, which outperforms existing state-of-the-art results.

Keywords: Retinal vessel segmentation · Fully convolutional network · Dense Adjacently Vessel Prediction · Separable Spatial and Channel Flow · Fundus image

1 Introduction

Retinal vascular network is the only vasculature which can be visualized and photographed in vivo. Retinal vascular imaging is able to provide clinically prognostic information for patients with specific cardiovascular and ophthalomologic diseases [1]. Segmenting out the retinal vessels is a prerequisite for monitoring the condition of retinal vascular network. Currently, retinal vessel segmentation highly relies on the manual work of experienced ophthalmologists, which is tedious, time-consuming, and of low reproducibility. As such, a fully-automated and accurate retinal vessel segmentation method is urgently needed to reduce the

© Springer Nature Switzerland AG 2019
H. Fu et al. (Eds.): OMIA 2019, LNCS 11855, pp. 112–120, 2019.
https://doi.org/10.1007/978-3-030-32956-3_14

workload on ophthalmologists and provide objective and precise measurements of retinal vascular abnormalities.

Several factors make this task challenging. The lengths and calibers of the vessels vary substantially from subject to subject. The presence of various types of lesions including hemorrhages, exudate, microaneurysm and fibrotic band can be confused with the vessels, so do the retinal boundaries, optic disk, as well as fovea. Furthermore, the relatively low contrast of the vessels and the low quality of some fundus images further increase the segmentation difficulties.

Numerous methods have been proposed for retinal vessel segmentation, both unsupervised and supervised. Unsupervised methods typically rely on mathematical morphology and matched filtering [2]. In supervised methods, ground truth data is used to train a classifier based on pre-identified features to classify each pixel into either vessel or background [3]. In the past few years, deep learning methods have seen an impressive number of applications in medical image segmentation, being able to learn sophisticated hierarchy of features in an end-to-end fashion. For example, Ronneberger et al. proposed U-Net to perform cell segmentation, which has become a baseline network for biomedical image segmentation, including retinal vessel segmentation [4]. Liskowski et al. used a deep neural network containing Structured Prediction trained on augmented retinal vessel datasets for retinal vessel segmentation [5]. Oliveria et al. combined the multi-scale analysis provided by Stationary Wavelet Transform with a fully convolutional network (FCN) to deal with variations of the vessel structure [6]. These approaches applying deep learning methods have significantly outperformed previous ones, achieving higher segmentation accuracies and computational efficiencies.

Despite their significant progress, existing deep learning approaches are facing the dilemma of effectively extracting vessels with small calibers versus maintaining high accuracy. Aforementioned approaches suffer from low capabilities of detecting thin vessels. Zhang et al. introduced an edge-aware mechanism by adding additional labels, which yielded a considerable improvement on predicting thin vessels but a decreased overall accuracy [7]. This is due to the fact that FCNs do not make use of spatial information in the pixel-wise prediction stage, but deploy a fully connected layer to each pixel separately. In such context, we propose a novel method for fundus image based retinal vessel segmentation utilizing a FCN together with *Separable Spatial and Channel Flow (SSCF)* and *Dense Adjacently Vessel Prediction (DAVP)* to capture maximum spatial correlations between vessels. Geometric transformations and overlapped patches are used at both training and prediction stages to effectively utilize the information learned at the training stage and refine the segmentation. Our method is quantitatively and qualitatively evaluated on the Digital Retinal Images for Vessel Extraction (DRIVE) [8] and Child Heart and Health Study in England (CHASE_DB1) datasets [9].

2 Method

2.1 Image Pre-processing

Although convolutional neural networks (CNNs) can effectively learn from raw image data, clear information and low noise enable CNNs to learn better. For a fundus image, its green channel is often used since it shows the best contrast in physiological structures with blood. And the blue channel contains relatively little physiological information. As such, simple but effective pre-processing is applied via $p_{i,j} = 0.25r_{i,j} + 0.75g_{i,j}$, where $p_{i,j}$ denotes the resulting pixel value at position (i,j) and $r_{i,j}$, $g_{i,j}$ respectively stand for the red channel value and the green channel value. Data augmentation is conducted to enlarge the training set by rotating each image patch by 90°, 180° and 270°.

2.2 Patch Extraction

Several studies have shown that CNNs can benefit from using overlapped patches extracted from large images for tasks that ignore contextual information [4, 6,7,11]. Given that the mean and variance of small image patches within a fundus image differ little, overlapped patches can also be applied to retinal vessel segmentation. Furthermore, a large amount of image patches can boost a CNN's performance by enlarging the sample size. In this work, a total of 1000 image patches of size 48×48 are randomly sampled from each training image. Center sampling is used and zero padding is performed if the center is located on image boundaries.

2.3 Fully Convolutional Network

FCNs can take input of arbitrary size and produce an output of the corresponding size with efficient inference and training by local connectivity and weight sharing [4]. FCNs typically have both down-sampling and up-sampling modules, which are used to respectively extract multi-scale features and reconstruct spatial information. In this work, we use U-Net as our baseline framework, which employs multiple skip connections to refine the detailed information lost in up-sampling modules [4]. As shown in Fig. 1, our overall architecture includes five stages. Extraction stage extracts low-level information from input images. Projection stage gradually projects multi-scale features into low-resolution feature maps that lie in high dimensional spaces. Mapping stage performs several non-linear mappings to explore more semantic information, providing guidance for pixels with low contrast and intensity. Refinement stage embeds spatial information into feature maps that have rich high-level information. By concatenating feature maps, this stage refines the semantic boundary. Reconstruction stage utilizes refined features to perform predictions, producing segmentation results.

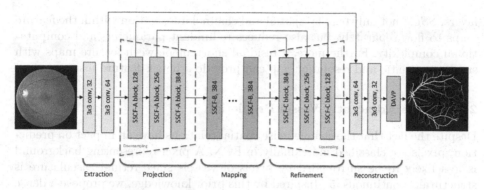

Fig. 1. Overall architecture of the proposed network. *SSCF-A*, *SSCF-B*, *SSCF-C* are detailed in Fig. 2. Please note, at Mapping stage, there can be any number of *SSCF-B* blocks, and we use two in our proposed network.

2.4 Separable Spatial and Channel Flow

Convolutional layers are designed to learn filters in a 3-dimension space (two spatial dimensions and one channel (grayscaled image intensity) dimension). Thus, a single convolution kernel should perform spatial and channel transformations jointly. However, filters in a single kernel usually conduct these two tasks implicitly, which may be vague and inefficient for high dimensional spaces. As such, we decouple the mappings of spatial correlations and cross-channel correlations sufficiently by factoring a kernel into a series of operations to perform those two mappings separately [10]. Specifically, we propose a block called *Separable Spatial and Channel Flow (SSCF)* and apply it to Projection, Mapping and Refinement stages, as shown in Fig. 2. Three depth-wise separable convolutional layers and one residual connection are contained in a *SSCF* block. Each depth-wise separable convolutional layer performs a depth-wise convolution followed by a point-wise convolution. A depth-wise convolution works as:

$$\mathbf{p}_{i,j,k} = \sum_{0 \le l,m < s} \mathbf{x}_{i+l-\lceil \frac{s}{2} \rceil, j+m-\lceil \frac{s}{2} \rceil, k} W_{i+l-\lceil \frac{s}{2} \rceil, j+m-\lceil \frac{s}{2} \rceil, k} \tag{1}$$

where $\mathbf{x}_{i,j,k}$, $\mathbf{p}_{i,j,k}$ respectively denote the input and result at position (i, j) and channel k, $W_{i,j,k}$ denotes the corresponding weight for $\mathbf{x}_{i,j,k}$ and s denotes the filter size. A point-wise convolution is written as:

$$\mathbf{y}_{i,j,c} = \sum_{k=1}^{C} \mathbf{p}_{i,j,k} W_{i,j,k,c} \tag{2}$$

where C denotes the number of input channels, $\mathbf{p}_{i,j,k}$ stands for the input at position (i, j) and channel k, $\mathbf{y}_{i,j,c}$ denotes the output at position (i, j) and channel c which can be any integer no larger than the total number of output channels. $W_{i,j,k,c}$ stands for the corresponding weight at position (i, j) for input channel k and output channel c. By stacking depth-wise separable convolutional layers, *SSCF* not only enables spatial and channel information within the feature maps to flow separately, but also reduces redundant parameters and computational complexity. Furthermore, a residual shortcut between feature maps with smaller semantic and resolution gap can provide a better feature fusion.

2.5   Dense Adjacently Vessel Prediction

Despite the fact that the channel-wise spatial relations have an impact on prediction, pixels are classified individually in FCN. A pixel representing background is less likely to be surrounded by pixels of vessels since retinal vasculature is structurally continuous [5]. Inspired by this prior knowledge, we propose a dense prediction cell named *Dense Adjacently Vessel Prediction (DAVP)*, as shown in Fig. 2. In addition to a 1×1 convolutional layer, an extra 5×5 convolution with non-linearity is introduced to filter redundant information. Then another 5×5 convolution utilizes spatial relations between pixels to perform prediction, refining the result from a 1×1 convolution branch via an element-wise addition.

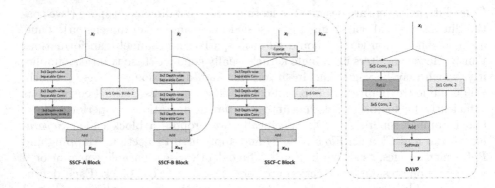

Fig. 2. Details of *SSCF* and *DAVP* employed in the proposed method.

3   Experiments

3.1   Datasets

We evaluated our method on the DRIVE and CHASE_DB1 public datasets. DRIVE consists of 40 fundus images of size 584×565 taken from both healthy adults and adults with mild diabetic retinopathy. There are 20 images for training and 20 images for testing [8]. CHASE_DB1 consists of 28 fundus images of size 999×960 taken from 14 10-year old children. For both datasets, gold standard segmentations are available. Since there is no official division into training and testing sets for CHASE_DB1 [9], we performed a 4-fold cross-validation in this case. The field of view masks for both datasets are publicly available [6], on which our quantitative evaluations are conducted.

3.2 Training

During the training process, Adam Optimizer was used to minimize cross entropy loss:

$$J(y, p) = -\sum_{k=1}^{C} y_k log(p_k) \tag{3}$$

where C refers to the number of classes, p and y respectively denote the probabilistic prediction and ground truth. The learning rate decayed by half for every 10 epochs, with an initial value of 0.001. The network was trained for 50 epochs, taking less than 1 h.

3.3 Implementation Details

The proposed method was implemented utilizing Keras with Tensorflow backend. All training and testing experiments were conducted on a workstation equipped with NVIDIA GTX Titan Xp.

3.4 Quantitative Results

To compare with other state-of-the-art results, we used four metrics for evaluation: accuracy (Acc), sensitivity (Sn), specificity (Sp) and area under the ROC curve (AUC-ROC). AUC-ROC is the key metric in retinal vessel segmentation considering the imbalance of classes. To obtain the binary vessel segmentation, a threshold of 0.5 is applied to the probability map.

Table 1 demonstrates the performance gains obtained from *SSCF* and *DAVP*, as evaluated on the DRIVE dataset. By decoupling the mappings of cross-channel correlations and spatial correlations, *SSCF* boosts the performance of U-Net, achieving an improvement on AUC-ROC by 0.09%. An incorporation of *DAVP* further improves the predictions by taking neighboring pixels into consideration during classification. These results imply *SSCF* and *DAVP* are helpful for embedding more spatial information between vessels. Figure 3 visualizes how *SSCF* and *DAVP* work.

Fig. 3. A test image patch from the DRIVE dataset. From left to right: one of the original image patches, the segmentation results from the baseline model, the baseline + *SSCF* model, the baseline + *SSCF* + *DAVP* model (the proposed) and the ground truth.

Table 1. Performance comparisons of different models on DRIVE.

Method	AUC-ROC
Baseline(U-Net)	0.9796
Baseline+preprocessing	0.9809
Baseline+preprocessing+$SSCF$	0.9818
Baseline+preprocessing+$SSCF$+$DAVP$	**0.9826**

We also compare our method with several other state-of-the-art methods in Tables 2 and 3. Our method outperforms all the other methods in terms of both accuracy and AUC-ROC. Also, reducing redundant parameters via the depth-wise separable convolutions strikingly shortens the training and inference time, making the proposed method being 90% faster than existing methods. Figure 4 shows representative segmentation results obtained from the proposed method on CHASE_DB1.

Table 2. Performance comparison with state-of-the-art methods on DRIVE, where t_T and t_I respectively denote time consumptions in training and inference.

Method	AUC-ROC	Acc	Sn	Sp	t_T	t_I
2nd observer [8]	N.A.	0.9473	0.7760	0.9725	N.A.	N.A.
Gu et al. [12]	0.9779	0.9545	**0.8309**	N.A.	N.A.	N.A.
Liskowski et al. [5]	0.9790	0.9535	0.7811	0.9807	8 h	92 s
Wu et al. [11]	0.9807	0.9567	0.7844	0.9817	16 h	10 s
Oliveria et al. [6]	0.9821	0.9576	0.8039	0.9804	N.A.	N.A.
Proposed method	**0.9826**	**0.9579**	0.7940	**0.9820**	1 h	1.3 s

Table 3. Performance comparison with state-of-the-art methods on CHASE_DB1.

Method	AUC-ROC	Acc	Sn	Sp
2nd observer [9]	N.A.	0.9560	0.7686	0.9779
Wu et al. [11]	0.9825	0.9637	0.7538	0.9847
Oliveria et al. [6]	0.9855	0.9653	0.7779	0.9864
Proposed method	**0.9865**	**0.9664**	**0.7878**	**0.9865**

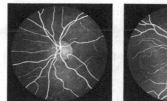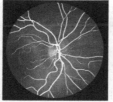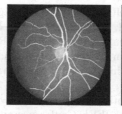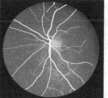

Fig. 4. Representative segmentation results superimposed on the original fundus images, obtained from the proposed method on CHASE_DB1.

4   Conclusion

In this paper, we proposed a novel FCN by incorporating *SSCF* and *DAVP* into U-Net for segmenting retinal vessels. The proposed *SSCF* and *DAVP* blocks can capture maximum spatial correlations between vessels, being able to solve the dilemma of maintaining high segmentation accuracy versus effectively extracting thin vessels. We demonstrated that the proposed method has state-of-the-art segmentation performance and high computational efficiency, which are essential in practical clinical applications. Future work will involve applying the proposed method to large-scale clinical studies.

Acknowledgement. This study was supported by the National Key R&D Program of China under Grant 2017YFC0112404 and the National Natural Science Foundation of China under Grant 81501546.

References

1. Patton, N., et al.: Retinal vascular image analysis as a potential screening tool for cerebrovascular disease: a rationale based on homology between cerebral and retinal microvasculatures. J. Anat. **206**(4), 319–348 (2005)
2. Hoover, A., Kouznetsova, V., Goldbaum, M.: Locating blood vessels in retinal images by piecewise threshold probing of a matched filter response. IEEE Trans. Med. Imaging **19**(3), 203–210 (2000)
3. Zhang, J., et al.: Retinal vessel delineation using a brain-inspired wavelet transform and random forest. Pattern Recogn. **69**, 107–123 (2017)
4. Ronneberger, O., Fischer, P., Brox, T.: U-Net: convolutional networks for biomedical image segmentation. In: Navab, N., Hornegger, J., Wells, W.M., Frangi, A.F. (eds.) MICCAI 2015. LNCS, vol. 9351, pp. 234–241. Springer, Cham (2015). https://doi.org/10.1007/978-3-319-24574-4_28
5. Liskowski, P., Krawiec, K.: Segmenting retinal blood vessels with deep neural networks. IEEE Trans. Med. Imaging **35**(11), 2369–2380 (2016)
6. Oliveira, A., Pereira, S., Silva, C.A.: Retinal vessel segmentation based on fully convolutional neural networks. Expert Syst. Appl. **112**, 229–242 (2018)
7. Zhang, Y., Chung, A.C.S.: Deep supervision with additional labels for retinal vessel segmentation task. In: Frangi, A.F., Schnabel, J.A., Davatzikos, C., Alberola-López, C., Fichtinger, G. (eds.) MICCAI 2018. LNCS, vol. 11071, pp. 83–91. Springer, Cham (2018). https://doi.org/10.1007/978-3-030-00934-2_10

8. Staal, J., et al.: Ridge-based vessel segmentation in color images of the retina. IEEE Trans. Med. Imaging **23**(4), 501–509 (2004)
9. Owen, C.G., Rudnicka, A.R., Mullen, R., Barman, S.A., et al.: Measuring retinal vessel tortuosity in 10-year-old children: validation of the computer-assisted image analysis of the retina (CAIAR) program. Invest. Ophthalmol. Vis. Sci. **50**(5), 2004–2010 (2009)
10. Chollet, F.: Xception: deep learning with depthwise separable convolutions. In: Proceedings of the IEEE Conference on CVPR, pp. 1251–1258 (2017)
11. Wu, Y., Xia, Y., Song, Y., Zhang, Y., Cai, W.: Multiscale network followed network model for retinal vessel segmentation. In: Frangi, A.F., Schnabel, J.A., Davatzikos, C., Alberola-López, C., Fichtinger, G. (eds.) MICCAI 2018. LNCS, vol. 11071, pp. 119–126. Springer, Cham (2018). https://doi.org/10.1007/978-3-030-00934-2_14
12. Gu, Z., et al.: CE-Net: context encoder network for 2D medical image segmentation. IEEE Trans. Med. Imaging (2019, in press)

Network Pruning for OCT Image Classification

Bing Yang[1,2], Yi Zhang[1(✉)], Jun Cheng[2(✉)], Jin Zhang[3], Lei Mou[2],
Huaying Hao[2], Yitian Zhao[2], and Jiang Liu[4]

[1] College of Computer Science, Sichuan University, Chengdu, China
`yi.zhang@scu.edu.cn`
[2] Cixi Institute of Biomedical Engineering, Ningbo Institute of Industrial
Technology, Chinese Academy of Sciences, Beijing, China
`chengjun@nimte.ac.cn`
[3] College of Electrical Engineering, Sichuan University, Chengdu, China
[4] Department of Computer Science and Engineering,
Southern University of Science and Technology, Shenzhen, China

Abstract. Convolutional neural network (CNN) has expanded rapidly,
and has been widely used in medical image classification. The large number of parameters in a neural network makes CNN models computationally expensive. This leads to slow inference speed, especially for 3D data
such as optical coherence tomography (OCT) for retinal images. A volume OCT scan of retina often contains hundreds of 2D images which
needs to be analyzed sequentially in a local computer with limited computational resources. We introduce network pruning to OCT images classification and propose an algorithm to prune networks. We compress the
popular classification models, such as ResNet and VGG. For example,
within 1% accuracy loss, we compress ResNet-18 from 44.8 MB to 69 KB
and VGG-16 from 537.1 MB to 194 KB. These pruned models are much
smaller and easier to deploy on the OCT devices. As for the inference
speed, the pruned models are 10 to 20 times faster than original models
for ResNet and VGG in CPU.

Keywords: Network pruning · OCT · Classification

1 Introduction

Optical Coherence Tomography(OCT) is a medical imaging modality developed
in the 1990s [6]. It is a high-resolution, non-invasive imaging technique which
has been widely used in the diagnosis and research of ocular diseases. Each
3D OCT scan often contains hundreds of 2D images. They are captured in a
low-performance computer. In recent years, deep learning is widely used in medical image analysis. Convolutional Neural Network(CNN) has proven to be very
effective in image processing. And it has also been adopted in OCT image classification [7,11,12]. The popular CNN models often require tens of hundreds of

© Springer Nature Switzerland AG 2019
H. Fu et al. (Eds.): OMIA 2019, LNCS 11855, pp. 121–129, 2019.
https://doi.org/10.1007/978-3-030-32956-3_15

megabytes for storage, and are regularly run on the GPU, thanks to its strong parallel computing power. But the storage and computing resource are constrained in some low-performance computers and mobile devices. Therefore, it becomes spontaneous to compress CNN models.

Over-parameterization is a widely recognized phenomenon of CNN [3,8]. In CNN model design, we first need to determine the model's architecture and the optimal amount of parameters. Both the architecture and parameter count decide the learning capability of CNNs. Large models have the ability to characterize complex problems, with the cost of greatly increased computation. However, insufficient parameters will restrict the learning ability. Over-parameterization often occurs when the selected model's parameterization is greater than what the task really needs. So, one natural idea is to train a large model and prune it by removing the redundant parameters.

The earliest work dates back to 1990 [2], which has been widely studied in recent years [1]. Generally, the computation complexity of CNNs is mainly related to the conv layers, and the number of parameters is dominated by the fully connected (FC) layers. Most pruning methods obey a three-stage pipeline: (a) training an original model. (b) prune the least important neural according to certain rules. (c) fine-tune to regain the lost performance. In the past few years, tremendous work has been made in the pruning rules. According to the granularity of the pruning, pruning methods can be commonly categorized into four groups [10]: fine-grained pruning (0-D), vector-level pruning (1-D), kernel-level pruning (2-D), and filter-level pruning (3-D). The latter three are also called structured pruning or coarse-grained pruning. It is pointed in [10]: different levels of pruning granularity often affect the acceleration efficiency for hardware and also affect the prediction accuracy. Coarse-grained methods are more friendly to hardware, but harder to maintain the accuracy than the fine-grained methods. In practice, the remaining weights are very likely to push training loss into a local minimum. Skilled network training is needed to avoid this situation. Very recently, Frankle and Carbin [4] propose a hypothesis that dense networks containing subnetworks can reach comparable accuracy in similar iterations. It prunes the smallest-magnitude weights and initializes the rest. Liu et al. [9] explore the value of network pruning. It comes to the conclusion that: the architecture of the pruned model is more valuable than the weights inherited from the original model.

Our work is partly inspired by [4] and [9]. This paper makes three contributions: (1) introduce network pruning to medical image classification for the first time, especially for OCT images; (2) different from [4], we prune networks on filter-level, because it is friendly to the current computing libraries and hardware; (3) we turn network pruning into the problem of finding optimal amount of parameters, and adopt Binary Search to accelerate the pruning process.

We get a large compression rate in model size on ResNet and VGG. Within 1% accuracy loss, for the channels percentage, we prune ResNet to 1%–2% and VGG to 1.7%–3.1%. As for the model size, we prune ResNet to 69 KB~ 228 KB,

VGG to 194 KB– 569 KB. And for inference time on CPU, we speed up ResNet more than 10x and VGG more than 20x.

2 Methods

We first make some changes to the traditional three-stage pipeline. We do not use any rules for sorting the importance of channels in conv layers, just prune the channels and initialize the rest, it can also be understood as replacing the old layer with a new layer, which has less channels and the weights are initialized. We use p% to represent the percentage for which the pruned model accounts for the original model. Then the task turns to be finding the percentage which is sufficient to achieve acceptable accuracy. We adopt Binary Search to quickly find the percentage. Our pruning strategy is shown in Algorithm 1.

Algorithm 1. Binary Search for Pruning Models

Input: original model, low: 0%, high: 100%, T: target accuracy
Output: pruned model
1: **while** $True$ **do**
2:     **if** low<high **then**
3:         $p\% \leftarrow middle = \frac{low+high}{2}$
4:         $temp\ model \leftarrow$ get $p\%$ of original model
5:         randomly initialize all weights of $temp\ model$
6:         $t \leftarrow$ accuracy = train $temp\ model$
7:         **if** $t \geq T$ **then**
8:             high$\leftarrow$ middle
9:         **else**
10:             low$\leftarrow$ middle
11:         **end if**
12:     **else**
13:         pruned model $\leftarrow temp\ model$
14:         break
15:     **end if**
16: **end while**

A short explanation for Algorithm 1 is given below:

1. First, we need to decide the original model and train it to get a satisfying accuracy. And set our target accuracy.
2. In the line 4, we get $p\%$ of every conv layer. The FC layer which always lies in the end needs to be prune at the same time.
3. In the line 5, we adopt the normal distribution which was used in [5] for the conv layer and FC layer initialization. We set all Batch Normalization layer's weight to be 1 and bias to be 0.

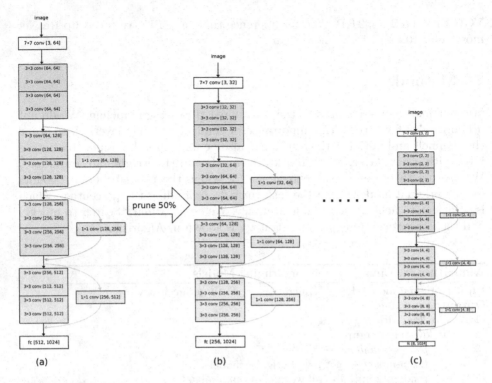

Fig. 1. Prune ResNet-18. (a) original model, (b) intermediate model during pruning, (c) the finally pruned model with 1.9% channels of original model

ResNet series is the stacking of residual modules, they have similar architecture, so we demonstrate our pruning strategy on the shallower model – ResNet-18, which is illustrated in Fig. 1.

According to the experience of model designing, the numbers of channels in conv layer will increase with the depth of network. When the first conv layer's channel is reduced to 1, it cannot be pruned anymore, but the next layers can still be pruned. The Binary Search is deployed in two speed:

– **Fast Speed:** Use Binary Search on every conv layer synchronously so as to find the small model quickly.
– **Slow Speed:** Skip a conv layer, and use Binary Search on the next conv layers.

The Binary Search accelerates the pruning process greatly. The traditional filter-level pruning algorithms prune a certain number of channels for each iteration, and the number is a hyperparameter that is decided by users. We use ResNet-18 as an example, the total number of channels is 4800. If we set the hyperparameter to be 100, and we reduce model from 100% to 3.1%, we have to train it 46 times, however, we only need to train it 5 times with Binary Search. As for the ResNet-101, the total number of channels is 52627, we train 514 times for traditional algorithm and 5 times for Binary Search.

3   Experiments

3.1   Dataset

We use the public dataset [7] in our experiments. We train models on this dataset. The OCT images are splitted into a training set of 83484 images, and a testing set of 1000 images. They are split into 4 categories: CNV, DME, DRUSEN, and NORMAL. The distribution is given in Table 1.

Table 1. Distribution for OCT images

Dataset	CNV	DME	DRUSEN	NORMAL	Total
Train	37,205	11,348	8,616	26,315	83,484
Test	250	250	250	250	1,000

3.2   Evaluation Methods

We compute Accuracy (Acc) to evaluate the pruned models. We compute the speedup ratio (SR) and model size compression ratio (CR) to represent the effect of compression. The equation is shown below.

$$Acc = \frac{TP + TN}{TP + FP + TN + FN} \tag{1}$$

$$SR = \frac{OT}{PT} \tag{2}$$

$$CR = \frac{OM}{PM} \tag{3}$$

In Eq. (1), TP, TN, FP and FN represent true positive, true negative, false positive and false negative. In Eq. (2), OT and PT represent inference time consuming of original model and pruned model. In Eq. (3), OM and PM represent the model size of original model and pruned model. We also compute Floating Point Operations Per Second(FLOPS), which is used to represent the computation complexity of CNN.

3.3   Experiment Settings

We train ResNet-18/34/50/101 and VGG-11/13/16/19 separately. The models are offered by PyTorch (v1.0.1). We train the models on a single NVIDIA GPU (GeForce GTX 1080 Ti). During the training phase, we adopt Stochastic Gradient Descent (SGD) as the optimizer, and we use cross-entropy to be the loss function. The initial learning rate is 0.01 for ResNet and 0.005 for VGG, with momentum is 0.9 and weight decay is 0.0005. We change the learning rate as the poly learning rate policy does in [14]. And the maximum epoch is 30 because the testing loss will converge before 30 epochs by experience. All images are resized to 224×224. We compute the inference time on the CPU (Intel Xeon Silver 4114).

4   Results

4.1   Accuracy

The accuracies of original models are over 99%, so we set 99% to be borderline, and explore the accuracy in different percentages of channels. It can be seen from Fig. 2: with the decreasing of models, firstly the accuracy fluctuates in a small range and then suffers a big loss in the same place. So, there is indeed a borderline for keeping accuracy, this percentage of parameters is suitable for classifying the OCT dataset.

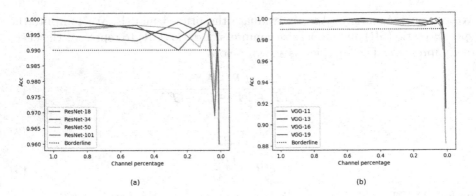

Fig. 2. Pruning process. Channel percentage represent the remaining channels account for the total channels in original model. (a) prune ResNet, Acc jumps off near 0.02. (b) prune VGG, Acc jumps near 0.03.

4.2   Model Size

The model size is important for deploying models on resource-constrained devices. The original models consume storage too much. Within 1% accuracy loss, we prune ResNet models to 1%–2% channel percentage. The details are shown in Table 2.

We can also generate two conclusions from Table 2:

Table 2. Model size of original and pruned models

Model	Original (MB)	p%	Pruned (KB)	CR
ResNet-18	44.8	1.9%	69	649.3
ResNet-34	85.3	1.5%	127	671.7
ResNet-50	99.4	1.6%	129	782.7
ResNet-101	170.6	1.8%	228	748.2
VGG-11	515.2	2.3%	307	1718.5
VGG-13	515.9	3.1%	524	1008.2
VGG-16	537.1	1.7%	194	2835.0
VGG-19	558.4	3.1%	569	1004.9

- For the same architecture, the more conv layers in original model, the bigger pruned model is. Because we just prune channels, not the entire layers.
- For different architectures, FC layer determines both original model size and pruned model size. The more, the bigger.

4.3 Parameters, FLOPS, Inference Time

The number of parameters, FLOPS, and inference time are important aspects to evaluate a model's performance. We test the models with one single OCT image on the CPU. The details are shown in Table 3. The pruned models are much smaller and faster than originals.

Table 3. Parameters, FLOPS, and inference time of different models

Model	Parameters (M)		FLOPS (G)		Inference time (S)		
	Original	Pruned	Original	Pruned	Original	Pruned	SR
ResNet-18	11.179	0.004	1.826	0.005	0.916	0.079	11.59x
ResNet-34	21.287	0.005	3.681	0.011	1.305	0.103	12.65x
ResNet-50	23.516	0.007	4.139	0.006	2.817	0.182	15.46x
ResNet-101	42.508	0.015	7.875	0.012	4.834	0.306	15.78x
VGG-11	128.783	0.071	8.567	0.006	2.813	0.103	27.19x
VGG-13	128.967	0.127	12.682	0.016	3.813	0.161	23.64x
VGG-16	134.277	0.042	17.308	0.006	5.251	0.162	32.46x
VGG-19	139.587	0.137	21.933	0.025	5.375	0.182	29.48x

4.4 Heatmaps

All of the above results indicate that: With 1% accuracy loss, we can get a very small model with our network pruning algorithm. But there is still a question:

Does the small model learn what the original model learns? We adopt Grad-CAM [13] to explain the prediction results of models. The heatmaps are shown in Fig. 3. The original and pruned models focus on the same area. The pruned model has a similar ability for feature extraction.

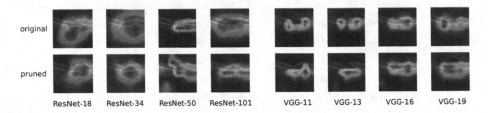

Fig. 3. Heatmaps of original and pruned models

5 Conclusion

For OCT images classification, there are huge redundant parameters in current popular models. Our network pruning algorithm can reduce them and get a smaller model that is suitable for computer with limited computational resources. And the model can be deployed on existing platforms directly. It will facilitate the application of CNNs on low performance and mobile devices.

Acknowledgment. This work was supported National Natural Science Foundation of China (61601029, 61602322), Grant of Ningbo 3315 Innovation Team, and China Association for Science and Technology (2016QNRC001).

References

1. Cheng, J., Wang, P., Li, G., Hu, Q., Lu, H.: Recent advances in efficient computation of deep convolutional neural networks. Front. Inf. Technol. Electron. Eng. **19**(1), 64–77 (2018)
2. LeCun, Y.: Optimal brain damage. In: International Conference on Neural Information Processing Systems (1989)
3. Denton, E., Zaremba, W., Bruna, J., LeCun, Y., Fergus, R.: Exploiting linear structure within convolutional networks for efficient evaluation. In: Proceedings of the 27th International Conference on Neural Information Processing Systems, NIPS 2014, vol. 1, pp. 1269–1277. MIT Press, Cambridge, MA, USA (2014). http://dl.acm.org/citation.cfm?id=2968826.2968968
4. Frankle, J., Carbin, M.: The lottery ticket hypothesis: finding sparse, trainable neural networks. In: ICLR (2019)
5. He, K., Zhang, X., Ren, S., Sun, J.: Delving deep into rectifiers: surpassing human-level performance on imagenet classification. In: Proceedings of the IEEE International Conference on Computer Vision 2015 Inter, pp. 1026–1034 (2015). https://doi.org/10.1109/ICCV.2015.123

6. Hee, M.R., et al.: Optical coherence tomography of the human retina. Arch. Ophthalmol. **113**(3), 325–332 (1995)
7. Kermany, D.S., et al.: Identifying medical diagnoses and treatable diseases by image-based deep learning. Cell **172**(5), 1122–1131.e9 (2018)
8. Ba, J., Caurana, R.: Do deep nets really need to be deep ? In: Advances in Neural Information Processing Systems, pp. 2654–2662 (2013)
9. Liu, Z., Sun, M., Zhou, T., Huang, G., Darrell, T.: Rethinking the value of network pruning. In: ICLR (2019)
10. Mao, H., et al.: Exploring the granularity of sparsity in convolutional neural networks. In: 2017 IEEE Conference on Computer Vision and Pattern Recognition Workshops (CVPRW), pp. 1927–1934, July 2017. https://doi.org/10.1109/CVPRW.2017.241
11. Perdomo, O., Otálora, S., González, F.A., Meriaudeau, F., Müller, H.: OCT-NET: a convolutional network for automatic classification of normal and diabetic macular edema using sd-oct volumes. In: 2018 IEEE 15th International Symposium on Biomedical Imaging (ISBI 2018), pp. 1423–1426, April 2018. https://doi.org/10.1109/ISBI.2018.8363839
12. Qiu, J., Sun, Y.: Self-supervised iterative refinement learning for macular OCT volumetric data classification. Comput. Biol. Med. 103327 (2019). https://doi.org/10.1016/j.compbiomed.2019.103327
13. Selvaraju, R.R., Cogswell, M., Das, A., Vedantam, R., Parikh, D., Batra, D.: Grad-CAM: visual explanations from deep networks via gradient-based localization. In: Proceedings of the IEEE International Conference on Computer Vision (2016). https://doi.org/10.1109/ICCV.2017.74
14. Zhao, H., Shi, J., Qi, X., Wang, X., Jia, J.: Pyramid scene parsing network. In: 2017 IEEE Conference on Computer Vision and Pattern Recognition (CVPR), pp. 6230–6239, July 2017. https://doi.org/10.1109/CVPR.2017.660

An Improved MPB-CNN Segmentation Method for Edema Area and Neurosensory Retinal Detachment in SD-OCT Images

Jian Fang[1], Yuhan Zhang[1], Keren Xie[2], Songtao Yuan[2], and Qiang Chen[1(✉)]

[1] School of Computer Science and Engineering,
Nanjing University of Science and Technology, Nanjing, China
chen2qiang@njust.edu.cn
[2] Department of Ophthalmology, First Affiliated Hospital with Nanjing
Medical University, Nanjing, China

Abstract. In the past, when facing with spectral-domain optical coherence tomography (SD-OCT) images of various types of age-related macular degeneration, such as neurosensory retinal detachment (NRD), pigment epithelial detachment (PED), hard exudate (HE), cystic edema (CE), and diffuse edema (DE), it was difficult to obtain satisfied segmentation results using traditional methods, because the DE and CE easily disturb the accuracy of NRD segmentation. In this paper, an improved multi-scale parallel branch convolutional neural network (MPB-CNN) network is proposed to perform the edema area (EA) segmentation and the NRD segmentation, where the supervised loss function is modified by adding area perimeter ratio constraint. The experiments on 98 cubes from 54 patients indicates that our method can achieve a mean overlap ratio 72.48% (NRD) and 75.93% (EA), respectively.

Keywords: Neurosensory retinal detachment · Edema area · MBP-CNN · SD-OCT · Semantic segmentation

1 Introduction

Central serous chorioretinopathy (CSCR) is a chronic disease, which is the leading cause of visual damage in middle-aged male. Neurosensory retinal detachment (NRD) is a prominent characteristic of CSC that occurs when subretinal fluid accumulates at the posterior pole [1]. Spectral-domain optical coherence tomography (SD-OCT) imaging technology can provide detailed characteristics of disease phenotypes, which has become an important imaging modality for the diagnosis of CSCR [2].

In the past decade, the traditional methods [3–5] have been used to segment the retinal edema lesions. Zheng *et al.* [3] proposed a fast segmentation algorithm with minimal expert's interaction to quantify intraretinal and subretinal fluid. Wang *et al.* [4] suggested a motion-estimation-based segmentation algorithm of fluid-associated

---

The first author of this manuscript is a master student.

© Springer Nature Switzerland AG 2019
H. Fu et al. (Eds.): OMIA 2019, LNCS 11855, pp. 130–138, 2019.
https://doi.org/10.1007/978-3-030-32956-3_16

regions. Wu *et al.* [5] put forward optical coherence tomography (OCT) fundus images to find abnormal locations and combined with fuzzy level sets for subretinal fluid segmentation.

Long *et al.* [6] advanced the fully convolutional networks (FCN) for pixels-to-pixels semantic segmentation for the first time. Ronneberger *et al.* [7] employed the U-Net, which included the shrink path of capturing context and the symmetric expanded path to accomplish precise location. Çiçek *et al.* [8] suggested the 3D U-Net, which extended the previous U-Net architecture by replacing all 2D operations with their 3D counterparts. Zhou *et al.* [9] presented a nested U-Net architecture (UNet++), essentially a deeply supervised encoder-decoder network, where the encoder and decoder subnetworks were connected through a series of nested and dense skip pathways. Zhang *et al.* [10] proposed the multi-scale parallel branch convolutional neural network (MPB-CNN), which was essentially three parallel branch networks are used for multi-scale feature extraction.

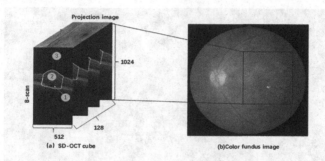

Fig. 1. SD-OCT volumetric images. (a) A SD-OCT cube, which contains $1024 \times 512 \times 128$ voxels with the corresponding clip size of 2 mm $\times$ 6 mm $\times$ 6 mm in (b). In (a), the yellow and red lines represent the boundaries of EA and NRD, respectively. The three numbers in (a) represent the three categories in our practical experiments (Color figure online).

This paper defined a novel lesion region, namely the edema area (EA), whose upper and lower boundaries are the internal limiting membrane and the Bruch's membrane (BM) respectively. To define the left and right boundaries of EA, we need to find the outermost point of edema between the retinal layers and make a line perpendicular to the BM. If the vertical line is outside the range of this image, that is drawn along the edge of the image. The motivation for defining EA is that the boundaries of diffuse edema are unclear, and doctors use the area for diagnosis in the clinic. Because the size of NRD in our dataset varies widely, multiscale features are capable of representing the NRD more robustly. The MPB-CNN is used to perform the EA and NRD segmentation. For further improving the segmentation results, the loss function of MPB-CNN was modified by adding an extra area perimeter ratio, which enables the network better for the complex edema.

132   J. Fang et al.

2 Methodology

2.1 Dataset

The data was collected from the Jiangsu Provincial People's Hospital using SD-OCT equipment (Carl Zeiss Meditec, Inc., Dublin, CA), including multiple cubes of the same patient at different times. As shown in Fig. 1(a), each cube includes $1024 \times 512 \times 128$ voxels, which corresponds to the trim size 2 mm $\times$ 6 mm $\times$ 6 mm. In the cube, each orientation slice is a B-Scan with the size 1024×512. Experts annotated the ground truth of EA and NRD manually. The complete data includes 98 cubes from 54 patients. Among these data, there were 6731 B-Scans with EA and 3620 B-Scans with NRD. The training set, validation set, and test set are composed of 70, 10, and 18 cubes, respectively.

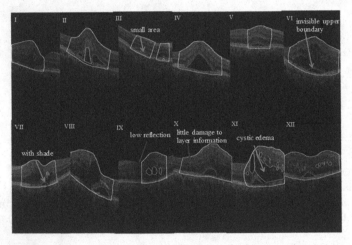

Fig. 2. Pathological analysis of edema. The area surrounded by the red curve is NRD, and the area surrounded by the yellow curve is the EA (EA contains NRD by default). The blue arrow indicates the influence factors for the NRD segmentation, and the purple arrow indicates the influence factors for the EA segmentation (Color figure online).

2.2 NRD and EA Pathologic Analysis

In retinal images, the purpose of semantic segmentation is to assign each pixel a specific label l, where $l \in \{1, 2, \ldots, K\}$, K is the number of classes. Figure 1 shows the EA and NRD segmentation, which is treated as 3 classes classifying problem.

Figure 2 shows the pathological features of the EA and NRD, which may include the following condition: (1) The upper boundary of NRD is not clear (VI). (2) Other retinal diseases can affect the NRD segmentation (XI). (e.g., cystic edema (CE), diffuse edema (DE), shadows caused by hard exudate (HE)). (3) The change of the NRD size is big (III). (4) There is the low reflection near EA (IX). (5) A part of the EA layer information is almost undamaged, and there are no serious lesions (X).

In this paper, the edema data contains a variety of data that is more complex than single lesion edema data. Some edemas are associated with other edema and cannot be completely distinguished by professional ophthalmologists. Several edema disturbances and some NRD boundaries are not clear. It is difficult for traditional machine learning to manually design appropriate features for the NRD and EA segmentation. However, deep learning has the ability to automatically extract classification features that avoid the limitations of manual features.

2.3 Network Architecture

Figure 3 shows the entire network architecture of MPB-CNN. Since the continuous B-scans of 3D SD-OCT cube have displacement during the image generation, our network is used for 2D B-scan instead of 3D cube. The three branches in MPB-CNN are parallel, and have the same structure, including encoder and decoder. We adopted three atrous convolution kernels with different rates to encode the input images. In the encoder section, we use a dense connection to capture the image information. In the decoder section, to improve the segmentation results we add the layer information of encoder into the deconvolutional layer. Figure 4 shows the network structure for each branch. Finally, the mapping of the low-level features and the three branches is cascaded achieved by a 3 × 3 convolution together as the output of the primary network to obtain the final segmentation result.

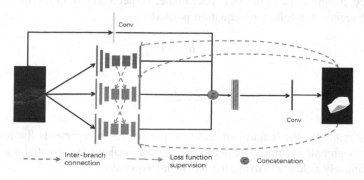

Fig. 3. The architecture of MPB-CNN. The three branches have the convolutional kernels with different scales to capture multi-scale information, and share the same network structure.

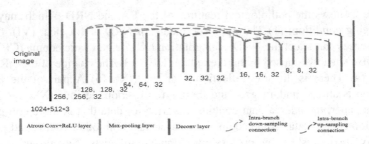

Fig. 4. The architecture of each branch. In the number below the convolution, the first two are the sizes of convolution, and the last one is the depth of convolution.

The MPB-CNN designed loss function that only focused on the gradient of the original graph boundary. In the case of complex edema adhesion, the MPB-CNN network is prone to misclassification. In order to solve this situation, this paper improved the loss function to reduce the area of the mistakenly segmented region by maximizing the area perimeter ratio.

We convert the prediction image and the ground truth into binary images. 0 represents the background, 1 represents EA as follows:

$$a_{i,j} = \begin{cases} 1 & \text{if } f_{i,j} \geq 1 \\ 0 & \text{if } f_{i,j} = 0 \end{cases} \tag{1}$$

$$a'_{i,j} = \begin{cases} 1 & \text{if } f'_{i,j} \geq 1 \\ 0 & \text{if } f'_{i,j} = 0 \end{cases} \tag{2}$$

where $f_{i,j}$, $f'_{i,j}$ are the values of the prediction label image and the ground truth in the (i,j) coordinate. $a_{i,j}$, $a'_{i,j}$ represent the values of the binaried prediction label image and the binaried ground truth in the (i,j) coordinate, respectively. To find the mistakenly segmented region, the following equation is used:

$$A_{i,j} = |a_{i,j} - a'_{i,j}| \tag{3}$$

$$S = \sum_i^x \sum_j^y A_{i,j} \tag{4}$$

where x, y represent the length and width of the image. S represents the area of the mistakenly segmented region. The perimeter of the mistakenly segmented region P can be approximately calculated using the following equation:

$$P = \sum_i^x \sum_j^y |A_{i,j} - A_{i-1,j}| \tag{5}$$

where $A_{i,j}$ represents the value of the mistakenly segmented region with coordinate (i,j).

We rewrote the optimization target as follows:

$$min_{\Theta}L(\Theta) \quad s.t. \, max\frac{S}{P} \tag{6}$$

Finally, the loss function for the final optimization can be rewritten as:

$$Loss = L(\Theta) + e^{-u*\left(\frac{S}{P}\right)} \tag{7}$$

where $L(\Theta)$ represents the final loss function in the MPB-CNN network, and the parameter u controls the proportion of $\left(\frac{S}{P}\right)$. We use $u = 0.001$ in Eq. (7) according to the constant experimental discovery.

2.4   Evaluation Design

To quantitatively evaluate the accuracy of the segmentation results, three evaluation criteria were used to evaluate the different segmentation methods: overlap ratio (Overlap), overestimated ratio (Overest), underestimated ratio (Undest) and pixel accuracy (Pixel Acc).

3   Results

The experiment was performed in a hardware condition with Intel Xeon CPU, one 11 GB NVIDIA GeForce GTX 1080 GPU and 64 GB RAM, and a software condition with Python 3.5 and Tensorflow. For patient independence, some patients were randomly selected for training and validation, and the remaining patients were used for testing. In order to verify our method, four deep learning methods (FCN [6], 3D-UNet [8] and UNet++ [9], MPB-CNN [10]) are compared.

Table 1 shows the NRD and EA segmentation accuracies of different methods, where our method achieved the overlaps 75.93% and 72.48% for EA and NRD, respectively. Our method is generally superior to other methods for the NRD segmentation and has improvement compared with the MPB-CNN for the EA segmentation.

Table 1. The performance comparison between different networks on edema dataset.

Lesion	Methods	FCN [7]	3D-UNet [9]	UNet++ [10]	MPB-CNN [11]	Our method
EA	Overlap [%]	0.7579	0.7722	0.7715	0.7515	**0.7593**
	Overest [%]	0.1525	0.1518	0.1671	**0.1323**	0.1481
	Undest [%]	0.0897	0.0760	**0.0614**	0.1162	0.0927
	Pixel Acc [%]	0.8931	0.9120	**0.9271**	0.8668	0.8936
NRD	Overlap [%]	0.6338	0.6630	0.6850	0.7084	**0.7248**
	Overest [%]	0.0466	0.0502	**0.0365**	0.0736	0.0502
	Undest [%]	0.3196	0.2868	0.2785	**0.2180**	0.2251
	Pixel Acc [%]	0.6673	0.7046	0.7116	0.7680	**0.7685**

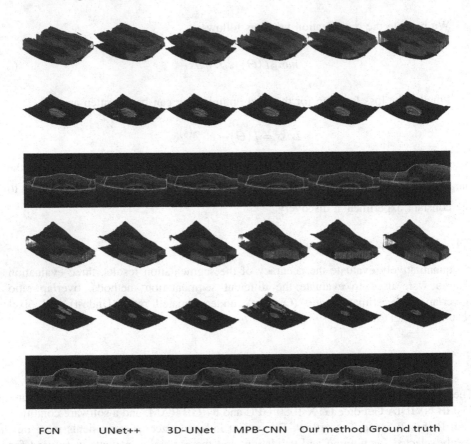

FCN   UNet++   3D-UNet   MPB-CNN   Our method Ground truth

Fig. 5. The 2D and 3D segmentation results of FCN [6], 3D-UNet [8], UNet++ [9], MPB-CNN [10] and our method. In the 1st, 2nd, 4th, and 5th rows, the dark blue surface is the BM layer, the red surface is the EA, and the green surface is the NRD. In the 3rd and 6th rows, the yellow and red lines indicate the ground truth of EA and NRD, respectively, and the green and blue lines indicates the automated segmentation of EA and NRD, respectively (Color figure online).

Figure 5 shows the segmentation comparison between our method and the other networks in 2D and 3D. In the case of a large number of cystic edema, our method has advantages, and our results are better than other results. In addition, the segmentation results indicate that the improved loss function is effective to reduce the segmentation error of cystic edema.

Figure 6 shows the several segmentation results of our method. We can find that our network can segment NRD and EA well compared with the ground truth except the last row. Our network model can adapt to the multi-scale NRD data and be able to accurately segment NRD against a variety of edema disturbances, such as PED, HE, CE and DE. The last row in Fig. 6 shows some failure results, where the large NRD containing HE and the NRD with weak boundary are difficult to be segmented accurately. These problems need to be solved in the future work.

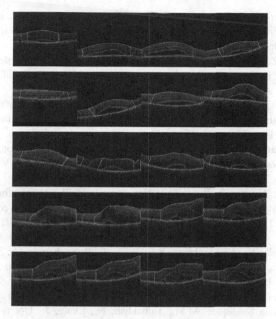

Fig. 6. Segmentation results of our method. The yellow line indicates the ground truth for the EA, and the green line indicates the automated segmentation of EA. The red line indicates the ground truth the NRD, and the blue line indicates the automated segmentation of the NRD. The last row shows the failure results (Color figure online).

4 Conclusion

It is difficult to obtain the satisfied NRD segmentation results using conventional methods for SD-OCT images, because the NRD is often disturbed by diffuse edema and cystic edema. Therefore, this paper intends to use deep learning to segment EA and NRD, and our main contribution is the improvement of loss function by adding an area perimeter ratio. Experimental results prove that our method is effective for the NRD and EA segmentation.

Acknowledgments. This work was supported by National Natural Science Foundation of China (61671242), Key R&D Program of Jiangsu Science and Technology Department (BE2018131), and Suzhou Industrial Innovation Project (SS201759).

References

1. Byon, I.S., Park, H.J., Park, S.W., et al.: Tissue layer image of the photoreceptor layer in central serous chorioretinopathy using SD-OCT. Ophthalmic Surg. Lasers Imaging Retina **43**(6), S16–S24 (2012)
2. Wang, J., Zhang, M., Pechauer, A.D., et al.: Automated volumetric segmentation of retinal fluid on optical coherence tomography. Biomed. Opt. Express **7**(4), 1577–1589 (2016)

3. Zheng, Y., Sahni, J., Campa, C., et al.: Computerized assessment of intraretinal and subretinal fluid regions in spectral-domain optical coherence tomography images of the retina. Am. J. Ophthalmol. **155**(2), 277–286 (2013)
4. Wang, T., Ji, Z., Sun, Q., et al.: Label propagation and higher-order constraint-based segmentation of fluid-associated regions in retinal SD-OCT images. Inf. Sci. **358**, 92–111 (2016)
5. Wu, M., Chen, Q., He, X.J., et al.: Automatic subretinal fluid segmentation of retinal SD-OCT images with neurosensory retinal detachment guided by enface fundus imaging. IEEE Trans. Biomed. Eng. **65**(1), 87–95 (2018)
6. Long, J., Shelhamer, E., Darrell, T.: Fully convolutional networks for semantic segmentation. In: Proceedings of the IEEE Conference on Computer Vision and Pattern Recognition, pp. 3431–3440 (2015)
7. Ronneberger, O., Fischer, P., Brox, T.: U-Net: convolutional networks for biomedical image segmentation. In: Navab, N., Hornegger, J., Wells, W.M., Frangi, A.F. (eds.) MICCAI 2015. LNCS, vol. 9351, pp. 234–241. Springer, Cham (2015). https://doi.org/10.1007/978-3-319-24574-4_28
8. Çiçek, Ö., Abdulkadir, A., Lienkamp, S.S., Brox, T., Ronneberger, O.: 3D U-Net: learning dense volumetric segmentation from sparse annotation. In: Ourselin, S., Joskowicz, L., Sabuncu, M.R., Unal, G., Wells, W. (eds.) MICCAI 2016. LNCS, vol. 9901, pp. 424–432. Springer, Cham (2016). https://doi.org/10.1007/978-3-319-46723-8_49
9. Zhou, Z., Rahman Siddiquee, M.M., Tajbakhsh, N., Liang, J.: UNet++: a nested U-Net architecture for medical image segmentation. In: Stoyanov, D., et al. (eds.) DLMIA/ML-CDS - 2018. LNCS, vol. 11045, pp. 3–11. Springer, Cham (2018). https://doi.org/10.1007/978-3-030-00889-5_1
10. Zhang, Y., Ji, Z., Wang, Y., et al.: MPB-CNN: a multi-scale parallel branch CNN for choroidal neovascularization segmentation in SD-OCT images. OSA Continuum **2**(3), 1011–1027 (2019)

Encoder-Decoder Attention Network for Lesion Segmentation of Diabetic Retinopathy

Shuanglang Feng[1], Weifang Zhu[1,2], Heming Zhao[1], Fei Shi[1],
Zuoyong Li[2], and Xinjian Chen[1,3(✉)]

[1] School of Electronics and Information Engineering, Soochow University,
Suzhou 215006, China
xjchen@suda.edu.cn

[2] Collaborative Innovation Center of IoT Industrialization and Intelligent
Production, Minjiang University, Fuzhou 350108, China

[3] State Key Laboratory of Radiation Medicine and Protection,
Soochow University, Suzhou 215123, China

Abstract. The segmentation of lesions such as retina edema, sub-retinal fluid and pigment epithelial detachment in optical coherence tomography (OCT) images is a crucial task for automated diagnosis of diabetic retinopathy. However, the multi-class lesion joint segmentation is very challenging due to the blurred boundary, complex structure, influence of noise, and the imbalanced class. In this paper, we propose a novel convolutional neural network with an encoder-decoder structure to perform joint segmentation of these three lesions. Unlike the common skip-connection employed in U-shape network for obtaining rich information from encoder feature map, we explore an encoder-decoder attention module (EDAM) via low-complexity non-local operation to capture more useful spatial dependency information between encoder feature and decoder feature. In this way, the network will take full advantage of the correlation information of the same stage feature and pay more attention to lesion areas. In order to capture large receptive fields and accurately segment small lesion, the modified light-weight residual network with dilated convolution is employed in encoding path. Besides, a hybrid loss, consisting of cross-entropy loss and multi-class Dice loss, is used to optimize our network. The proposed method was evaluated on a public database: AI-challenger 2018 for automated segmentation of retinal edema lesions, and achieved a compelling performance with less parameters compared to state-of-the-art networks.

1 Introduction

Diabetic Retinopathy (DR) is one of the main blinding diseases, affecting the normal life of approximately 34% of diabetic patients. DR may cause many symptoms that appear on the retina such as retina edema (RE), sub-retinal fluid (SRF), pigment

---

S. Feng, and W. Zhu—These authors contributed equally to this work.

H. Fu et al. (Eds.): OMIA 2019, LNCS 11855, pp. 139–147, 2019.
https://doi.org/10.1007/978-3-030-32956-3_17

epithelial detachment (PED). Optical coherence tomography (OCT) images are widely used in ophthalmology clinic for the diagnosis of retinal diseases. Therefore, the automatic segmentation of lesions in OCT images plays a key role in the diagnosis and treatment of DR. However, the main challenge for this task lies in the following factors: (1) Joint segmentation of multiple types of lesions is difficult due to the extreme imbalance of the data distribution between different lesions. (2) The boundary of the retina edema area (REA) is blurred and difficult to determine. (3) The influence of speckle noise and vascular artifacts is severe.

In recent years, many segmentation studies on DR lesion have been proposed. Most of these methods such as graph search based methods [1, 2], kernel regression based methods [3] have two stages: retinal layer segmentation, lesion delineation. The computational bottleneck caused by algorithm optimization makes it urgent to develop an end-to-end solution. Recently, many deep learning methods based on convolutional neural networks (CNN) [4] have been applied to medical image analysis. Guha et al. [5] proposed a ReLayNet with position indices pooling for retinal layer and fluid pocket segmentation. Freerk et al. [6] utilized typical U-shape neural network for segmentation of macular edema. Most of these CNN-based approaches only focus on single type lesion. To the best of our current knowledge, there are no methods based on CNN for joint lesion segmentation in DR OCT images, which is always challenging to jointly segment imbalanced medical data for CNN based on encoder-decoder architecture. In this paper, we design a novel and efficient network to address these problems.

The skip-connection of U-Net [4] is an ingenious design, which can combine the encoder feature to make up for the information loss caused by downsampling. However, simple skip-connection ignores contextual information and is an indiscriminate combination of semantic information that will introduce noise of irrelevant clutters. Previous work [7, 8] overlooked this important detail. Although [9, 10] proposed a global convolutional network (GCN) between encoder and decoder, it can't capture the global information in the real sense and ignore the spatial correlation. General non-local model (NLM) [11] was applied in video classification and semantic segmentation, which utilized a self-attention mechanism to get the approximate autocorrelation information. In this paper, a novel encoder-decoder attention module (EDAM) based on non-local operation is employed to generate approximate cross-correlation information between encoder feature and decoder feature. In this way, the network can enhance the correlated responses of focused object and weaken the uncorrelated responses in global view through a controllable information flow from encoder. Furthermore, for non-local operation, we explore a low-complexity representation to handle high computational complexity issue. Besides, in order to obtain the high-resolution feature map and accurately segment small lesion, we improve a lightweight residual network [12] with dilated convolution [13] as backbone network to extract feature and employ a hybrid loss consisting of cross-entropy loss and multi-class Dice loss to alleviate the imbalanced data problem.

Consequently, our main contributions include: (1) An efficient encoder-decoder attention network is proposed for joint lesion segmentation in DR OCT images. (2) The proposed encoder-decoder attention module (EDAM) can capture richer global features and model spatial correlation between encoder feature and decoder feature. (3) We achieve impressive results with less parameters compared to state-of-the-art networks on public database: AI-challenger 2018 for automated segmentation of retinal edema lesions.

2 Method

2.1 Proposed Network Architecture

Figure 1 is an overview architecture of our proposed encoder-decoder attention network for joint segmentation of three DR lesions (REA, SRF, PED). In order not to lose the information of small lesions during the downsampling, we improve a residual network with dilated convolution as the encoder to extract high resolution feature map. The dilated convolution with rate of 2 is employed in block3 and rate of 4 is employed in block4 like [14]. Therefore the output size of feature map from encoder is 1/8 of input image. For the convenience of skip-connection, the first downsampling is performed by a 3 × 3 convolution layer with a stride of 2 after a 7 × 7 convolution layer and a bottleneck layer [12]. The next two downsampling layers are in the first bottleneck layer of block1 and block2 respectively. Note that the channel expansion rate is set to 2. In decoder part, bilinear interpolation operation is applied in three upsampling layers to quickly restore the original image size. The boundary refinement (BR) blocks [9] are used to refine the edges of the feature map, which consist of two convolution layers with residual design. It is worth noting that we employ an EDAM between each corresponding stage of encoder and decoder to capture more correlated information about prediction feature map from encoder path.

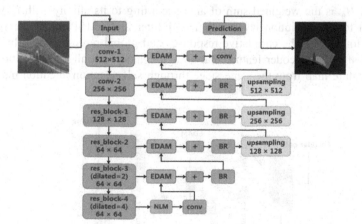

Fig. 1. An overview of our proposed network architecture. EDAM and NLM represent encoder-decoder attention module, general non-local module, respectively. BR and conv represent boundary refinement block and convolutional layer, respectively.

2.2   Encoder-Decoder Attention Module

Attention mechanism is widely used in natural language processing (NLP) [15] and computer vision (CV) [16] field, which can draw global information and obtain rich feature. In this work, we propose an encoder-decoder attention module (EDAM) shown in Fig. 2. To model the spatial correlation over the global view between encoder feature and decoder feature via non-local operation. Here, the feature $\mathbf{E} \in \mathbb{R}^{C \times H \times W}$ from encoder path generates two feature maps $\mathbf{V}$ and $\mathbf{K}$ via two convolution layers with 1×1 filters, respectively, where $\{\mathbf{V}, \mathbf{K}\} \in \mathbb{R}^{C' \times H \times W}$. Meanwhile, the feature map $\mathbf{Q} \in \mathbb{R}^{C' \times H \times W}$ is generated by $\mathbf{D} \in \mathbb{R}^{C' \times H \times W}$ from decoder path through the convolution layer with 1×1 filter. C' is the channel number of feature map $\mathbf{D}$, which is less than C. Then we reshape $\mathbf{V}, \mathbf{K}$ and $\mathbf{Q}$ to $\mathbb{R}^{C' \times N}$, where $N = H \times W$. After that, we use a hyperparameter factor $\alpha = \sqrt{C' \times H \times W}$ to normalize the result of matrix multiplication between the transpose of $\mathbf{K}$ and $\mathbf{Q}$ and generate the pixel-wise affinity attention map $\mathbf{A} \in \mathbb{R}^{N \times N}$:

$$\mathbf{A} = \frac{\mathbf{K}^{\mathrm{T}} \mathbf{Q}}{\alpha} \tag{1}$$

Then we perform matrix multiplication between $\mathbf{V}$ and attention map $\mathbf{A}$ to obtain a final feature map $\mathbf{H}$ and reshape it to $\mathbb{R}^{C' \times H \times W}$, here, $\mathbf{H}$ is feature map that has been weighted by correlated contextual information between encoder feature map and decoder feature map:

$$H_{ic} = \sum_{j=0}^{N} A_{ji} V_{jc} \ \{i, j \in [1, 2, \dots N], \ c \in [1, 2, \dots C']\} \tag{2}$$

Where H_{ic} is the weighted sum of all j according to its affinity with i. After this, $\mathbf{H}$ is input to a convolution layer with 1×1 filter and added to decoder path local feature $\mathbf{D}$ to enhance the correlative responses and global representation. In this way, any position in the encoder feature map is aggregated with all other positions in the decoder feature map from the same stage through self-adaption attention maps.

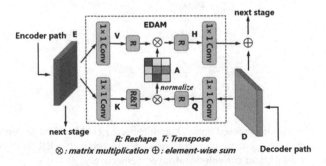

Fig. 2. The detail of encoder-decoder attention module

2.3   Low-Complexity Representation

Representing the relationship between any two pixels requires a complex matrix multiplication operation to obtain a huge attention map, its complexity is $\mathcal{O}(N^2)$ in both time and space, where $N = H \times W$ indicates the spatial dimension of feature map. Because of the high resolution of feature map in semantic segmentation task, we cannot afford for directly implementing EDAM with our limited GPU memory. Fortunately, we explore an efficient and alternative method to achieve the same target with the associative law of matrix multiplication. Note that we obtain the final feature map $\mathbf{H}$ through two matrix multiplication operations in Sect. 2.2.

$$\mathbf{H} = \frac{\mathbf{V}(\mathbf{K}^{T}\mathbf{Q})}{\alpha} = \frac{(\mathbf{V}\mathbf{K}^{T})\mathbf{Q}}{\alpha} \tag{3}$$

According to the associative law, we could perform $\mathbf{Z} = \mathbf{V}\mathbf{K}^{T} : \mathbb{R}^{C' \times N} \times \mathbb{R}^{N \times C'} \rightarrow \mathbb{R}^{C' \times C'}$ first and then calculate $\mathbf{Z}\mathbf{Q}$, where C' is channel of decoder path feature map and much smaller than N. These operations greatly reduce the time and space complexity from $\mathcal{O}(N^2)$ to $\mathcal{O}(C'^2)$, and can be easily embedding into other encoder-decoder network. We apply this compatibility representation to each stage between encoder and decoder to replace the original skip-connection except for the bottom of network which employs a NLM.

2.4   Loss Function

To alleviate the problem that Dice loss is sensitive to small structures or absent classes, we employ a hybrid loss consisting of cross-entropy loss and multi-class Dice loss to preform joint lesion segmentation. The total loss can be expressed as:

$$\mathcal{L} = \mathcal{L}_{Dice} + \lambda \mathcal{L}_{CE}$$
$$= 1 - \frac{1}{C} \sum_{c=0}^{C-1} \frac{2 \sum g * p + \varepsilon}{\sum (g+p) + \varepsilon} - \lambda \frac{1}{N} \sum g \, \log(p) \tag{4}$$

Where λ is a weighted coefficient between Dice loss $\mathcal{L}_{Dice}$ and cross-entropy loss $\mathcal{L}_{CE}$ and set to 1 in our work, $p \in [0, 1]$ is the predicted probability, $g \in \{0, 1\}$ is the true label, N is the spatial dimension of feature map, C is a sum of the total number of lesion classes and one background class, and ε is a small smoothing factor.

3   Experiments

Databases: We evaluated our proposed network in a public database: AI-challenger 2018 for automated segmentation of retinal edema lesions, which contains of 100 cubes with the size of $1024 \times 512 \times 128$ and is annotated manually by experts. All of these cubes contain REA, but PED and SRF only involve some cubes and are surrounded by REA. Since the annotation problem, 68 cubes was chosen for training and 30 cubes was chosen to averagely divided into Part A and Part B for testing. In the databases, background accounts for 92.94% of all the voxels, REA, SRF and PED takes up 6.19%, 0.84% and 0.03%, respectively, which indicates the segmentation of SRF and PED will suffer from the extremely imbalanced data distribution problem. Besides, the 3D OCT cube is a huge burden for limited GPU memory. To address these problems, we extract 2D B-scan image as the input image of segmentation task.

Patch Extraction and Data Augmentation: The size of 2D B-scan slice is 1024×512. According to our statistics, the lesion area always lie in a square of size $l \times l$, where $l = 512$. Therefore, for the slice containing the lesion, we randomly cropped a $l \times l$ patch which include whole lesion, and for the slice with only background, a $l \times l$ patch was randomly cropped on single slice. After that a single slice containing PED was reapplied to random cropping 9 times to balance data. Now we got a more balanced training data than before with about 12300 patches. In the inference phase, we applied a sliding window strategy with the window size $l \times l$ to take tiles, and the stride is $l/4$. To alleviate the border effects in segmentation task, we put more weight on the middle of prediction feature map during the splicing process.

Implementation: We employed Keras and tensorflow to implement our proposed method. The optimizer was stochastic gradient descent (SGD) with the "poly" learning rate policy. The basic learning rate and momentum were set to 0.01 and 0.9, respectively. The batch size and epoch were set to 12 and 40, respectively. For data augmentation, we only applied online random left-right flipping. All of these were preformed on three NVIDIA Tesla K40 GPUs with 12 GB memory.

4 Evaluation and Results

We compared feature maps from common skip-connection and EDAM. Figure 3 displays the visualization of feature maps, in which shows that common skip-connection introduces unnecessary noise and has no any discrimination for feature. However, our EDAM can enhance the correlated responses of focused object via modeling long-range spatial dependencies in global view and suppress irrelevant noise.

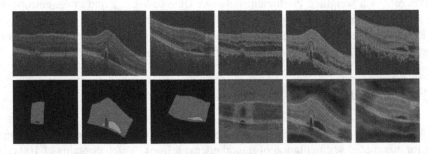

Fig. 3. The visualization of feature maps. Left top: original patch. Left bottom: ground truth. Right top: feature map from common skip-connection. Right bottom: feature map from EDAM.

We carried out comprehensive experiments on Part A and Part B OCT cubes with the measure of average Dice scores in Table 1. In order to evaluate the effectiveness of our proposed EDAM, an ablation experiment was conducted. Baseline represents our network employing the common skip-connections, EDAM represents encoder-decoder attention module inserted in each stages, we observe that it yields the more gain when EDAM is inserted into network comparing with Baseline network. Moreover, there is only a little increase of parameters because EDAM is inserted in relatively shallow

Table 1. The performance of different segmentation models and our proposed method on Part A and Part B OCT cubes, measured with average global Dice scores (±standard deviation).

Method	Part A Dice (%)				Part B Dice (%)				#Para
	REA	SRF	PED	Ave	REA	SRF	PED	Ave	
U-Net	73.23 (±14.6)	60.96 (±37.3)	41.35 (±38.7)	58.51	75.25 (±13.6)	60.13 (±38.1)	28.59 (±26.2)	54.65	31.04 M
Attention U-Net [18]	74.56 (±14.8)	62.07 (±36.7)	40.94 (±35.3)	59.19	76.53 (±16.5)	61.34 (±35.9)	29.67 (±23.4)	55.84	29.82 M
PSPNet [14]	75.99 (±14.3)	59.42 (±38.4)	42.76 (±36.7)	59.39	77.58 (±14.1)	59.42 (±37.3)	31.69 (±25.4)	56.23	46.77 M
Res-FCN [10]	74.15 (±16.0)	62.88 (±35.9)	43.50 (±37.22)	60.17	76.83 (±13.4)	61.24 (±36.3)	29.51 (±24.4)	55.86	37.05 M
V-Net [17]	76.23 (±15.2)	57.33 (±36.2)	39.48 (±36.24)	57.68	77.04 (±12.7)	58.45 (±38.0)	26.16 (±23.1)	53.88	45.60 M
Baseline	75.58 (±14.3)	64.67 (±35.7)	39.59 (±35.2)	59.95	76.25 (±14.6)	62.14 (±36.7)	31.82 (±24.2)	56.74	19.79 M
Baseline + EDAM	**76.51** (**±14.2**)	**67.75** (**±35.2**)	**45.14** (**±34.9**)	**63.13**	**78.04** (**±13.1**)	**64.83** (**±37.4**)	**32.01** (**±23.6**)	**58.29**	19.87 M

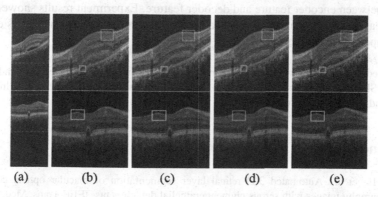

(a)   (b)   (c)   (d)   (e)

Fig. 4. Examples of segmentation results. (a) Original OCT B-scans, (b) Ground truth, (c) The segmentation results of Baseline + EDAM, (d) The segmentation result of Baseline, (e) The segmentation result of Res-FCN [12]. Note that red area, green area and blue area represent REA, SFR and PED, respectively (Color figure online).

layers with a few channels. We also report the performance of our proposed network compared with other state-of-the-art segmentation methods. To be fair, we employed the same hybrid loss to these comparable methods. It can be seen that all the Dice scores of REA, SRF, PED and the average Dice score of our method were superior to all the other methods both on Part A and Part B. Especially, the number of parameters in our network (20 M) is only 65% of U-Net (31 M), while it achieved a dramatic improvement of 6% in average Dice scores. And the similar phenomena has been also shown in other results. Our method also achieved a significant improvement (student's t-test, p-value = 0.04) on SRF segmentation with a 3%–5% Dice increase to the second-best one. Due to the false positive prediction of cube without SRF and PED, the variance is relatively large. All these performances indicated that our network is more efficient than all compared methods, which benefits from EDAM capturing global correlation information between encoder and decoder.

Two examples of segmentation results are shown in Fig. 4. In the first row, the size of PED areas is too small to Res-FCN, while our method could segment it precisely. In addition, for relatively large REA lesion, our method achieved a very similar result to the ground truth, which benefited from EDAM can aggregate the correlative information between encoder and decoder in global view. It can be also seen that the proposed method can eliminate more false positives than Res-FCN method from second row. Overall, the encoder-decoder attention network is capable of the accurate and effective segmentation for DR joint lesion due to its strong global correlative information aggregation ability.

5 Conclusion

We proposed a novel network with encoder-decoder attention module for the multi-class joint lesion segmentation of diabetic retinopathy. The proposed method models the long-range spatial dependencies and captures more correlative contextual information between encoder feature and decoder feature. Experiment results showed that it has a great potential for imbalanced data medical image segmentation with its efficient and compelling performance.

Acknowledgments. This work was supported by the National Natural Science Foundation of China (NSFC) (61622114, 81401472) and Collaborative Innovation Center of IoT Industrialization and Intelligent Production, Minjiang University (No. IIC1702).

References

1. Shi, F., et al.: Automated 3-D retinal layer segmentation of macular optical coherence tomography images with serous pigment epithelial detachments. IEEE Trans. Med. Imaging **34**(2), 441–452 (2015)
2. Sun, Z., et al.: An automated framework for 3D serous pigment epithelium detachment segmentation in SD-OCT images. Sci. Rep. **6**, 21739 (2016)
3. Chiu, S.J., et al.: Kernel regression based segmentation of optical coherence tomography images with diabetic macular edema. Biomed. Opt. Express **6**(4), 1172–1194 (2015)

4. Ronneberger, O., Fischer, P., Brox, T.: U-Net: convolutional networks for biomedical image segmentation. In: Navab, N., Hornegger, J., Wells, W.M., Frangi, A.F. (eds.) MICCAI 2015. LNCS, vol. 9351, pp. 234–241. Springer, Cham (2015). https://doi.org/10.1007/978-3-319-24574-4_28
5. Roy, A.G., et al.: ReLayNet: retinal layer and fluid segmentation of macular optical coherence tomography using fully convolutional networks. BOE **8**(8), 3627–3642 (2017)
6. Venhuizen, F.G., et al.: Deep learning approach for the detection and quantification of intraretinal cystoid fluid in multivendor optical coherence tomography. Biomed. Opt. Express **9**(4), 1545–1569 (2018)
7. Badrinarayanan, V., et al.: Segnet: a deep convolutional encoder-decoder architecture for image segmentation. IEEE Trans. PAMI **39**(12), 2481–2495 (2017)
8. Jégou, S., et al.: The one hundred layers tiramisu: fully convolutional densenets for semantic segmentation. In: CVPR Workshop, pp. 11–19 (2017)
9. Peng, C., Zhang, X., Yu, G., Luo, G., Sun, J.: Large kernel matters–improve semantic segmentation by global convolutional network. In: CVPR, pp. 4353–4361 (2017)
10. Liu, Z., et al.: Towards clinical diagnosis: automated stroke lesion segmentation on multi-spectral MR image using convolutional neural network. IEEE Access **6**, 57006–57016 (2018)
11. Wang, X., Girshick, R., Gupta, A., He, K.: Non-local neural networks. In: CVPR, pp. 7794–7803 (2018)
12. He, K., Zhang, X., Ren, S., Sun, J.: Deep residual learning for image recognition. In: CVPR, pp. 770–778 (2015)
13. Yu, F., Koltun, V.: Multi-scale context aggregation by dilated convolutions. arXiv preprint. arXiv:1511.07122 (2015)
14. Zhao, H., Shi, J., Qi, X., Wang, X., Jia, J.: Pyramid scene parsing network. In: CVPR, pp. 2881–2890 (2017)
15. Vaswani, A., et al.: Attention is all you need. In: NIPS, pp. 5998–6008 (2017)
16. Hu, J., et al.: Squeeze-and-excitation networks. In: CVPR, pp. 7132–7141 (2018)
17. Milletari, F., et al.: V-net: fully convolutional neural networks for volumetric medical image segmentation. In: Fourth International Conference on 3D Vision, pp. 565–571 (2016)
18. Oktay, O., et al.: Attention U-Net: learning where to look for the pancreas. arXiv preprint. arXiv:1804.03999 (2018)

Multi-discriminator Generative Adversarial Networks for Improved Thin Retinal Vessel Segmentation

Gabriel Tjio[1]([⊠]), Shaohua Li[1], Xinxing Xu[1], Daniel Shu Wei Ting[2], Yong Liu[1], and Rick Siow Mong Goh[1]

[1] Institute of High Performance Computing, A*STAR, Singapore, Singapore
gabriel-tjio@ihpc.a-star.edu.sg
[2] Singapore Eye Research Institute, Singapore, Singapore

Abstract. Retinal vessel segmentation is an important step in clinical analysis of fundus images. Low contrast and the imbalanced pixel ratios between thick and thin vessels make accurate segmentation of the thin vasculature extremely challenging. In this paper, we present a novel multiscale segmentation method named Multiple discriminator generative adversarial network (MuGAN). MuGAN contains multiple discriminators with different effective receptive fields, which are sensitive to features at different scales. These discriminators jointly teach the segmentation (generator) network to pay attention to multiscale patterns. In addition, multiple discriminators allow our model to incorporate multiple inputs, such as edge enhanced vessel images, during training. We evaluated our method on the publicly available DRIVE and STARE datasets. MuGAN achieved an overall area under the Receiver Operator Characteristic Curve (AUC) of 0.979 for DRIVE and 0.981 for the STARE dataset. On segmenting thin retinal vessels, MuGAN showed quantitative and qualitative improvements on baselines.

Keywords: Multiscale segmentation · Generative Adversarial Network · Fundus imaging

1 Introduction

Retinal vessels have been shown to be important features for cardiovascular disease detection [1]. Additionally, changes in retinal vessel diameters are associated with the progression of retinal diseases [2] and higher risk of cardiovascular mortality [3]. Accordingly, accurate segmentation of retinal vessels plays an important role in characterizing the patient's eye and cardiovascular health.

Recent work on retinal vessel segmentation has focused on deep learning approaches. However, accurate segmentation of thin retinal vessels remains challenging. Retinal vessel widths can range from 1 pixel to 10 pixels in diameter, making multiscale segmentation necessary for good performance. As noted in [4], large retinal vessels tend to be more accurately segmented than thin vessels because (1) the majority of vessel pixels belong to thick vessels, and (2) thick vessels typically have higher contrast than thin vessels.

© Springer Nature Switzerland AG 2019
H. Fu et al. (Eds.): OMIA 2019, LNCS 11855, pp. 148–155, 2019.
https://doi.org/10.1007/978-3-030-32956-3_18

This paper aims to address this problem with the following framework. We propose a novel multiscale, multi-input generative adversarial network (MuGAN) for retinal vessel segmentation. GANs [5] have been used in broad applications including image segmentation [6]. Son et al. [7] applied GANs on retinal vasculature segmentation. While [7] manifests good performance, it is not explicitly designed to detect thin vascular details. This work is inspired by the multiple discriminators approach [8], which suggests that a group of limited capacity discriminators can better capture the probability distribution of the training data compared to a single discriminator.

We build upon their work by using multiple discriminators with different architecture for improved multiscale segmentation. Firstly, we use convolutional layers of varying dilation rates for different discriminators. This preserves the resolution of the extracted features while varying the effective receptive field. Yu et al. [9] implemented dilated convolutions to combine multiscale features. We therefore suggest that introducing convolutions with different dilation rates will enable the discriminator to capture additional high level features. Additionally, multiple discriminators allow the GAN to incorporate different inputs: (1) the multiscale features extracted using dilated convolutional layers, and (2) the postprocessed outputs from the generator network and the postprocessed ground truth. We also use edge-enhanced ground truth images as additional input for training because the enhanced vessel boundaries mitigate the imbalance between thick and thin vessel pixels (Fig. 1). Finally, we use skeletal metrics based on [10] for performance evaluation because commonly used performance metrics such as AUC (Area under the receiver operator characteristic curve) may not adequately reflect the thin vessel segmentation accuracy.

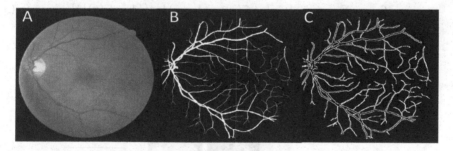

Fig. 1. Example of (A) retinal fundus image, (B) ground truth vessel segmentation, (C) edge enhanced ground truth image. The edge enhanced ground truth increases the visibility of the thin vessel branches and mitigates the imbalance between thin and thick vessel pixels.

To assess MuGAN segmentation accuracy, we evaluated MuGAN, MUGAN$_{noedge}$ (MuGAN without edge enhanced inputs), GAN$_{single}$ (single generator discriminator pair) and other approaches [11, 12] on the publicly available datasets DRIVE [13] and STARE [14].

2 Methods

2.1 Generative Adversarial Networks for Retinal Vessel Image Segmentation

In Generative Adversarial Networks, the generator and discriminator are alternatively trained to minimize and maximize the objective function. The standard objective function for GAN is as follows:

$$\min_{\theta_G} \max_{\theta_D} \mathcal{L}(D, G) = \mathbb{E}_{x \sim p_{data(x)}} \log(D(x)) + \mathbb{E}_{z \sim p_{data(z)}} \log(1 - D(G(z))) \qquad (1)$$

where θ_D and θ_G refer to the parameters for the discriminator D and generator G respectively, x: source data, z: noise inputs, p_{data}: probability distribution of data. The generator must be able to generate an output such that the discriminator is unable to differentiate between that output and the ground truth. We modify the objective function [7] to include the loss function for the a given discriminators D_i, λ is used to weight the segmentation loss $\mathcal{L}_{seg}$:

$$\min_{\theta_G} \max_{\theta_{D_i}} \mathcal{L}(D_i, G) + \lambda \mathcal{L}_{seg}(G) \qquad (2)$$

The segmentation loss function $\mathcal{L}_{seg}$ uses the binary cross entropy loss, which compares the generator output $G(x)$ with the ground truth (segmented vasculature) y and x is the source data (input fundus images):

$$\mathcal{L}_{seg} = \mathbb{E}_{x,y \sim p_{data(x)}} \{-y \log G(x) - (1 - y) \log(\{1 - G(x)\})\} \qquad (3)$$

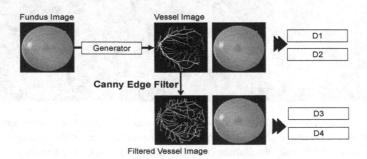

Fig. 2. Workflow for training the multiple discriminator GAN (MuGAN). The discriminators D1 and D3 have the same architecture. Similarly, D2 and D4 have the same architecture. D1 and D2 (D3 and D4) have different effective receptive field sizes.

Figure 2 illustrates the proposed workflow. The discriminators D1 (D3) and D2 (D4) have different effective receptive field sizes. We vary effective receptive field size by changing the dilation rate of the convolutional layers [9]. The first set of discriminators (D1 and D2) was trained to distinguish between the ground truth and the trained segmentations. The second group of discriminators (D3 and D4) was trained to distinguish between edge enhanced ground truth and edge enhanced segmented vessels. The ground truth vessel images were processed with a Canny edge filter before training with the default settings (lower bound for thresholding: 10% of maximum pixel value of input; upper bound of maximum pixel value: 20%; σ of Gaussian: 1.0), instead of tuning the parameters to suit the training data. This was done because the default settings are likely to have the best performance across a wide range of image types. The generated vessel images were filtered before training the discriminator. Edge enhancement was not applied during training, eliminating the need for a differentiable edge detection method.

2.2 Architecture

Figure 3 describes the MuGAN architecture. The basic unit of the generator and discriminator architecture is the convolutional block, which comprises of a Conv2D 3 × 3, n (n is the depth: 32, 64, 128), a batch normalization layer, an activation function and a max pooling layer of kernel size 3 × 3, The Rectified Linear Unit (ReLu) activation function was used for all convolution layers, with the exception of the last layer (Conv2D 1 × 1, 1) in the generator, which uses a sigmoid activation function. The generator is based on the UNET architecture [15] for its capability to resolve high level and low level features. The discriminators have relatively shallow architectures (3 convolutional layers) to minimize computational requirements. For the discriminator, skip connections are used to pass the output from each convolutional block to be subsequently concatenated and pooled using a global max pooling layer. The discriminator D2 has the same architecture as D1, except that D2 has a dilation rate of 2 for the convolutional block. D1 and D3 both do not utilize dilated convolutional filters (dilation rate = 0). D3 has the same architecture as D1 while D4 has the same architecture as D2.

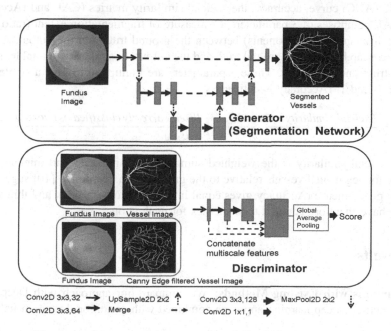

Fig. 3. Architecture of the generator and discriminator networks.

3   Experiments

3.1   Dataset

We use the publicly available DRIVE [13] and STARE [14] dataset. The DRIVE dataset contains 40 images from diabetic patients (584 × 565 pixels, 45° field of view FOV). The STARE dataset contains 20 images (605 × 700 pixels, 35° FOV). The DRIVE dataset is evenly split for training and testing. We performed leave one out cross validation (LOOCV) for the STARE dataset. Training is performed with 19 images and testing is conducted on the 'left out' image. 20 iterations of these train-test cycles are performed to evaluate performance for all images. The fundus images and ground truth segmented vessel images are padded to 640 × 640 pixels (DRIVE) or 720 × 720 pixels (STARE). Data augmentation was performed by flipping and rotating the images, generating 266 images (DRIVE) and 252 images (STARE).

3.2   Training

We implement our approach in Python 2.7 using the Keras framework. We used the Adam optimizer, initial learning rate 2×10^{-4} and trained for 20 epochs. It took approximately 2–3 h to train the proposed MuGAN model on the DRIVE dataset and each fold of the STARE dataset (Intel(R) Xeon(R) W-2145 CPU, NVIDIA Titan Xp GPU, Keras version 2.2.4).

3.3   Evaluation Criteria

We employ the following evaluation criteria: area under the Receiver Operator Characteristic (AUC) curve, accuracy, the skeletal similarity metrics (CAL and rAccuracy) [10]. CAL comprises of 3 parameters, C: measure of fragmentation extent (ratio of the number of connected components) between the ground truth and the output, A: measure of overlap between the ground truth and the output, L: similarity in total lengths of ground truth and the output. These 3 parameters are multiplied to give a single score between 0 and 1. rAccuracy is defined as

$$rAcc = \frac{Skeletal\ Similarity \times Pixels_{vessels} + Pixels\ correctly\ classified\ as\ background}{Pixels_{total}} \quad (4)$$

where skeletal similarity is the weighted sum of curve similarity and thickness consistency for segmented vessels relative to the ground truth. Yan et al. [10] suggest that their proposed metric rAccuracy gives equal importance to both thick and thin vessels and we have therefore adopted their metrics for evaluation.

4   Results

Comparisons with Existing Methods: We compare our approach with DeepVessel [11], which is a deep learning approach combined with conditional random fields and

M2U-Net [12], an efficient deep learning approach based on the UNET architecture [15]. To further ascertain the effectiveness of including multiscale inputs and edge information, we train the GAN networks (GAN$_{single}$ and MuGAN$_{noedge}$). Both networks do not use the edge enhanced vessel images for training (GAN$_{single}$ comprises of the generator and discriminator D1; MuGAN$_{noedge}$ comprises of the generator, discriminators D1 and D2). Tables 1 and 2 gives the accuracy, CAL, rAccuracy and AUC for the DRIVE and STARE datasets.

The results show that our method performs better than other methods, particularly for CAL and rAccuracy. MuGAN$_{noedge}$ shows mixed performance relative to GAN$_{single}$, performing better for the STARE dataset than the DRIVE dataset. GAN$_{single}$ shows comparable performance with M2U-Net [12] on DRIVE. M2U-Net [12] was not trained on STARE and therefore not evaluated for STARE. DV [11] is also comparable with GAN$_{single}$ on STARE, but also performs less well than MuGAN. MuGAN has higher accuracy compared to MuGAN$_{noedge}$ and GAN$_{single}$. These results suggest that edge information, combined with multiscale discriminators, help improve segmentation accuracy. Though the edge information used in this study emphasizes the boundaries at the expense of eroding the center of the vessels, the results suggest that segmentation accuracy is not worsened due to the loss of information.

Table 1. Averaged performance metrics for DRIVE test images (n = 20)

	DV [11]	M2UNET [12]	GAN$_{single}$	MuGAN$_{noedge}$	MuGAN
Accuracy	0.949	0.951	0.953	0.953	**0.955**
CAL	0.687	0.812	0.811	0.786	**0.829**
rAccuracy	0.898	0.931	0.939	0.932	**0.944**
AUC	–	0.971	0.976	0.975	**0.978**

Table 2. Performance metrics for STARE images (n = 20, Leave-one-out cross validation)

	DV [11]	GAN$_{single}$	MuGAN$_{noedge}$	MuGAN
Accuracy	0.958	0.958	0.959	**0.960**
CAL	0.713	0.715	0.737	**0.746**
rAccuracy	0.942	0.950	0.956	**0.960**
AUC	–	0.974	0.978	**0.981**

Cross Training: We also perform cross training between the STARE and DRIVE datasets. Table 3 shows the cross training performance. Overall, performance across the different approaches is similar. MuGAN AUC fell from 0.981 to 0.966 and 0.978 to 0.953 for the STARE and DRIVE datasets. MuGAN$_{noedge}$ AUC fell from 0.978 to 0.968 and 0.975 to 0.956 for the STARE and DRIVE datasets. One possible explanation for the results is that STARE contains more pathological images than DRIVE, resulting in lower performance when the model trained on DRIVE is implemented on the STARE dataset. This is also supported by the greater difference in CAL scores (DRIVE: 0.829 to 0.641, STARE: 0.746 to 7.61) for MuGAN. Interestingly, MuGAN$_{noedge}$ performs better than GAN$_{single}$ for both datasets. We suggest that the multiple discriminators reduce the effects of overfitting. Figure 4 also shows the qualitative improvements with our approach, with MuGAN detecting the thin vasculature absent in the outputs from other methods.

Table 3. Cross training performance

	train (DRIVE), test (STARE)			train (STARE), test (DRIVE)		
	GAN_{single}	$MuGAN_{noedge}$	MUGAN	GAN_{single}	$MuGAN_{noedge}$	MUGAN
Accuracy	0.946	0.949	**0.949**	0.947	**0.950**	0.949
CAL	0.599	0.640	**0.641**	0.736	**0.764**	0.761
rAccuracy	0.939	0.943	**0.946**	0.918	0.924	**0.928**
AUC	0.951	**0.956**	0.953	0.964	**0.968**	0.966

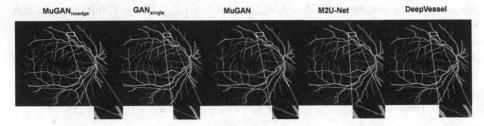

Fig. 4. Difference images obtained from comparing the ground truth with the segmented results. Red indicates vessels incorrectly classified as background, blue indicates background wrongly classified as vessels and green indicates correctly classified vessel pixels. The rAccuracy for this subject is 94.9%, 95.8% and 95.9% for $MuGAN_{noedge}$, GAN_{single} and MuGAN respectively, while M2U-Net [12] and DeepVessel [11] achieve a rAccuracy of 94.9% and 90.5%. (Color figure online)

5 Conclusion

Segmentation of thin retinal vessels is one of the main challenges in retinal image analysis. We implemented a multiple discriminator GAN approach to improve multi-scale segmentation of the retina vessels, with particular focus on the thin vessels. Our proposed method has two main novel aspects: multi-input multiscale discriminators which use (1) discriminators with varying effective receptive field sizes and (2) additional input (edge filtered vessel images) to improve segmentation performance. Future work will explore custom loss functions for thin vessel segmentation and other approaches to convey edge information during training.

References

1. Poplin, R., et al.: Prediction of cardiovascular risk factors from retinal fundus photographs via deep learning. Nat. Biomed. Eng. **2**, 158–164 (2018)
2. Klein, R., et al.: The relation of retinal vessel caliber to the incidence and progression of diabetic retinopathy: XIX: the Wisconsin epidemiologic study of diabetic retinopathy. Arch. Ophthalmol. **122**, 76–83 (2004)

3. Wang, J.J., et al.: Retinal vessel diameter and cardiovascular mortality: pooled data analysis from two older populations. Eur. Heart J. **28**, 1984–1992 (2007)
4. Yan, Z., Yang, X., Cheng, K.T.: A three-stage deep learning model for accurate retinal vessel segmentation. IEEE J. Biomed. Health Inform. **23**(4) (2019). https://ieeexplore.ieee.org/document/8476171
5. Goodfellow, I.J., et al.: Generative Adversarial Networks. arXiv:1406.2661 (2014)
6. Luc, P., Couprie, C., Chintala, S., Verbeek, J.: Semantic Segmentation using Adversarial Networks. arXiv:1611.08408 (2016)
7. Son, J., Park, S.J., Jung, K.-H.: Retinal Vessel Segmentation in Fundoscopic Images with Generative Adversarial Networks. arXiv:1706.09318 (2017)
8. Durugkar, I., Gemp, I., Mahadevan, S.: Generative Multi-Adversarial Networks. arXiv:1611.01673 (2016)
9. Yu, F., Koltun, V.: Multi-Scale Context Aggregation by Dilated Convolutions. arXiv:1511.07122 (2015)
10. Yan, Z., Yang, X., Cheng, K.: A skeletal similarity metric for quality evaluation of retinal vessel segmentation. IEEE Trans. Med. Imaging **37**, 1045–1057 (2018)
11. Fu, H., Xu, Y., Lin, S., Kee Wong, D.W., Liu, J.: DeepVessel: retinal vessel segmentation via deep learning and conditional random field. In: Ourselin, S., Joskowicz, L., Sabuncu, Mert R., Unal, G., Wells, W. (eds.) MICCAI 2016. LNCS, vol. 9901, pp. 132–139. Springer, Cham (2016). https://doi.org/10.1007/978-3-319-46723-8_16
12. Laibacher, T., Weyde, T., Jalali, S.: M2U-Net: Effective and Efficient Retinal Vessel Segmentation for Resource-Constrained Environments. arXiv:1811.07738 (2018)
13. Staal, J., Abramoff, M.D., Niemeijer, M., Viergever, M.A., Ginneken, B.V.: Ridge-based vessel segmentation in color images of the retina. IEEE Trans. Med. Imaging **23**, 501–509 (2004)
14. Hoover, A.D., Kouznetsova, V., Goldbaum, M.: Locating blood vessels in retinal images by piecewise threshold probing of a matched filter response. IEEE Trans. Med. Imaging **19**, 203–210 (2000)
15. Ronneberger, O., Fischer, P., Brox, T.: U-Net: convolutional networks for biomedical image segmentation. In: Navab, N., Hornegger, J., Wells, William M., Frangi, Alejandro F. (eds.) MICCAI 2015. LNCS, vol. 9351, pp. 234–241. Springer, Cham (2015). https://doi.org/10.1007/978-3-319-24574-4_28

Fovea Localization in Fundus Photographs by Faster R-CNN with Physiological Prior

Jun Wu[1], Jiapei Wang[1], Jie Xu[2(✉)], Yiting Wang[1], Kaiwei Wang[1],
Zongjiang Shang[1], Dayong Ding[3], Xirong Li[4,5], Gang Yang[5], Xuemin Jin[6],
Yanting Wang[6], Fangfang Dai[6], and Jianping Fan[7]

[1] School of Electronics and Information, Northwestern Polytechnical University,
Xi'an, China
[2] Beijing Tongren Hospital, Capital Medical University,
Beijing Ophthalmology and Visual Science Key Lab, Beijing, China
fionahsu920@foxmail.com
[3] Vistel AI Lab, Visionary Intelligence Ltd., Beijing, China
[4] School of Information Science and Technology,
University of Science and Technology of China, Hefei, China
[5] School of Information, Renmin University of China, Beijing, China
[6] Henan Provincial Peoples' Hospital, Zhengzhou, China
[7] University of North Carolina-Charlotte (UNCC), Charlotte, NC, USA

Abstract. A macular fovea is a physiological structure of the human retina, which is an essential optical center. The distance between the lesion area and the foveal center determines the severity degree of visual impacts. Therefore, accurate fovea localization is the basis of the computer-aided ophthalmic diagnosis and vision screening. A simple but effective fovea localization algorithm based on the Faster R-CNN and physiological structure prior is presented. First, a fovea localization model and an optic disc localization model are trained separately. Then, for each fundus photograph, both candidate areas of the fovea and the location of the optic disc are predicted using two pre-trained models. Next, prior knowledge of the physiological adjacent relationship between a fovea and an optic disc is applied to eliminate unreasonable candidate bounding boxes. Finally, the ultimate bounding box of the fovea is determined by the best candidate. Experiments were conducted on a private dataset with 5,203 fundus photographs and the public Messidor dataset including 1200 fundus photographs. The accuracy of the foveal location in the offset scale of 1/2 optic disc diameter on the Messidor is 99.58%, which is 0.71% higher than the state-of-the-art (98.87%).

Keywords: Fovea localization · Faster R-CNN · Fundus photography · Optic disc · Physiological prior · Messidor

© Springer Nature Switzerland AG 2019
H. Fu et al. (Eds.): OMIA 2019, LNCS 11855, pp. 156–164, 2019.
https://doi.org/10.1007/978-3-030-32956-3_19

1 Introduction

In a human retina, the typical tissues are vessels, an optic disc, a macula, *etc.* An optic disc is a vertical oval in the nasal side. A macula is an oval-shaped pigmented area near the retinal center. A fovea is located in the macular center and it is a small pit that contains the largest concentration of cone cells, which is the most sensitive part to light. Once any macular lesion occurs, the vision will be significantly affected. Fast and accurate automatic fovea detection can greatly improve the diagnostic efficiency of retinal diseases, and it also can provide a basis for the large-scale screening of common retinal diseases.

A foveal center is located within an inferior macula without any vascular distribution. As shown in Fig. 1(a), in a color fundus photograph, the fovea is the lowest brightness area in the macular center, roughly located on a symmetrical axis which divides the upper and lower branches of the entire vascular network. From the optic-disc center to its temporal direction, a foveal center is located around 2–2.5 optic disc diameter (*dd*) away, with a small horizontal angle [15].

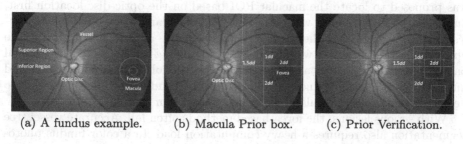

(a) A fundus example.	(b) Macula Prior box.	(c) Prior Verification.

Fig. 1. Prior physiological relationships. (a) Illustration of a macular, a fovea, and an optic disc in a fundus photograph. (b) A left eye example for macular restriction box. (c) Verification using Physiological Relationship Prior. (Color figure online)

By using different structure information, the fovea localization methods have three categories. One way is to locate the fovea using the appearance characteristics of a fovea itself [7,11,15]. Singh et al. [15] first proposed a pre-processing method to improve the image contrast between the macula and its surrounding areas in the red channel and then located the foveal center by searching the darkest area. Lu et al. [7] proposed a linear metric operator based on the brightness and shape features of the macula, which can search from different directions and reflect the gray gradient changes. The image enhancement by an adaptive manifold filter and mathematical morphological operations were applied to locate the fovea [11].

The second way is to use contexts such as the surrounding vessels or optic disc [2,5,10,18]. Niemeijer et al. [10] proposed a cost function based on global and local cues to find the correct foveal center. Zheng et al. [18] first segmented a vessel network and obtained the macular Region of Interest (ROI). Then, a circular region was fitted along vessel endings around the macula to locate the fovea. In [5], based on the optic-disc location, a one-dimensional scanned intensity profile analysis was applied for the fovea localization. Dashtbozorg *et al.* [2] presented an innovative super-elliptical filter to localize the optic disc and the fovea simultaneously.

The last way is to combine the fovea, the vessel network and the optic disc together [1,4,6,8,13,14,16,19]. In [16], an approximate macular position was obtained by using the optic-disc position, where the foveal center was estimated as the lowest response point from a directional matched filter. In [13], the main vessels were extracted first, and the macular ROI was obtained from the vessel tree. Then the foveal center was determined at a distance below the temporal direction of the optic disc. Gegundez-Arias et al. [4] utilized the information of an optic disc and a vascular tree to obtain an approximate location of the macula. In [1] visual and anatomical feature-based criteria were combined with respect to the optic disc and the vascular tree. Kao et al. [6] localized the optic-disc center first by the template matching, and determined the disc-fovea axis by searching the vessel-free region. Finally, the fovea center was detected by matching the fovea template around the center of the axis. In [19], another method was proposed to locate the macular ROI based on the optic-disc location first, and then the foveal center was determined by macular features and mathematical morphology. In [14], the vessel segmentation was applied first to locate the optic disc, and the fovea was localized by matching the expected directional pattern. Molina-Casado et al. [8] proposed a methodology combining intra- and inter-structure relational knowledge based on the candidate tuples validation to detect the optic disc, macula and vascular network in a unified framework.

With serious lesions, the fovea-based methods often fail. Accurate vessel tree segmentation also requires a heavy computation load. In a color fundus photograph, the macular characteristics are not obvious and accurate macular boundary is often difficult to distinguish. However, an optic disc is relatively brighter, which benefits its localization. Besides, retinal lesions have relatively less influence on locating an optic disc rather than that of a macula. A hemorrhage, macular edema, and other retinal lesions affect the macular appearance tremendously or even cover it completely leading to fovea locating failures. In this case, using the optic-disc position to locate the macula is the only option.

In this paper, we propose a simple but effective fovea localization algorithm by the Faster R-CNN deep learning framework with the prior structure relationship between a fovea and an optic disc. Our main contributions include: (1) The prior physiological relationship between a fovea and an optic disc in the human retina is sufficiently explored to eliminate unreasonable foveal candidates. (2) This prior knowledge can also help to build a hypothetic candidate in case of complete failure on the fovea localization, especially for retinal diseases with severe lesions, which is also difficult for a professional ophthalmologist. (3) The foveal annotation is supposed to be only a center point, not the regional bounding boxes. As a result, this case is more difficult than the case with accurate size information of the bounding boxes at the same time. The localization algorithm needs to determine the optimal size of the bounding boxes that will be fed into the Faster R-CNN network for training.

2   The Proposed Method

Our main objective is to automatically locate the foveal center in color fundus photographs. First, a bounding box of a fovea based on its center annotated by the ophthalmologists is fed to train a Faster R-CNN model. Then, the preferred candidate foveal areas of the top N rank predicted by the pre-trained Faster R-CNN model are screened as the foveal candidate areas. Finally, the candidates with the top N highest scores are further verified by using the prior physiological relationship between a fovea and an optic disc in the human retina, resulting in the final foveal center as the best output (Fig. 2).

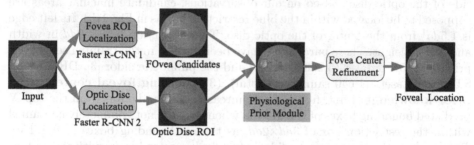

Fig. 2. Flowchart of our proposed fovea localization algorithm.

2.1   Optic Disc Localization

An optic disc provides useful information to locate the fovea according to their physiological relationships. A Faster R-CNN [12] deep learning network is applied to locate the optic disc first. Based on our evaluations, the predicted location of an optic disc in a high accuracy can be directly considered as its real center.

2.2   Fovea Localization

(1) Basic Network: The Faster R-CNN [12] is popular to locate the bounding box of an image object. It is a combination of the Fast R-CNN and the region proposal networks (RPN). We use it as our baseline network.

(2) Optimizing the Size of the Foveal Bounding Box for Training: To locate a fovea, the rough macular region should be detected first. A progressive macular area with vascular distribution exists around a macula, which also helps for the fovea localization. As a result, it is necessary to determine a suitable size of the bounding box for the target fovea to feed into the Faster R-CNN for training. Different sizes of foveal bounding boxes, which contain different context characteristics, affect the performance of the fovea localization. As a result, generally, its size can be optimized on the evaluation set. We try different box size for fovea localization model training from 120×120, 140×140, ..., to 240×240 in pixels, and finally 200×200 is verified as the best option.

2.3 Verification Using Physiological Relationship Prior

In the case that a macula can be roughly located successfully, the position of the optic disc can be applied to refine and optimize the foveal localization, where candidate bounding boxes are with top N highest scores from the Faster R-CNN prediction. The detailed steps are: (1) **Judging left or right eye.** In a typical fundus photograph of the posterior pole, the position of the optic disc is located in the nasal side. Taking the middle vertical line, if the predicted center of the optic disc is on the left side of the midline, it is a left eye and the macula is on the right side of the optic disc. Conversely, it is judged as the right eye, and the macular is on the left side of the optic disc. (2) **Determining macular candidates.** As shown in Fig. 1(b), an optic disc is on the left of the middle vertical line, and it is judged as a left eye. The macula is located in the right-down side of the optic disc. Based on our observations, candidate macular areas are supposed to be located within the blue restriction box as in Fig. 1(b). Its left edge is $1.5dd^1$ from the center of the optic disc. This restriction box is $2dd$ in width and $3dd$ in height. These parameters have been verified in the training sets of our private MF5K dataset (5,203 samples) and the public Messidor [3], DRIVE and STARE datasets (1,640 samples in total). (3) **Locating foveal center using prior knowledge.** First, filtering out unreasonable candidates from the top-N predicted bounding boxes of the macula when they are not completely contained within the restriction box of $3dd \times 2dd$, as the red bounding boxes in Fig. 1(c). The one from the remaining candidates (marked as green boxes) with the highest score from the Faster R-CNN is determined as a final predicted foveal bounding box. (4) **Failure Case Correction.** The prior relationship of $2.5dd$ from the center of the optic disc also helps to build a hypothetic candidate in the extreme case of complete failure on fovea localization with the severe lesions.

3 Evaluation

3.1 Experimental Setup

We validate our proposed method on two datasets. (1) **Messidor** [3]: a public dataset with 1,200 color fundus photographs of the posterior pole are acquired by 3 ophthalmologic departments. The foveal center annotations containing the pixel size of the disc diameter (dd) in each fundus photograph, which are provided by Niemeijer et al. [10], are applied in our evaluation. (2) **MF5K:** A private dataset with 5,203 fundus photographs is collected from a local partner hospital. The ophthalmologists manually annotate foveal centers, and the majority of photographs include diseases such as diabetic retinopathy (DR) in around 97%. Therefore, it is a very challenging dataset to locate the fovea due to the interference of the DR disease. In addition, the training set, validation set, and test set are partitioned randomly by 50%, 25%, and 25% respectively, resulting in 2,601, 1,301 and 1,301 fundus photographs for each subset.

---

[1] Empirically, dd is about 80 pixels in a 500×500 fundus photograph [17].

The experiments are conducted using a Pytorch deep learning framework under Ubuntu 14.04 with an NVIDIA 1080Ti GPU. The images in the MF5K are resized to 500×500 pixels. A carefully-verified 200×200 macular region is used for training the Faster R-CNN model. The shared convolution layer uses the VGG16 to extract features, scales are set as [8, 16, 32], ratios as [0.5, 1, 2], the momentum in the optimizer as 0.9, and the learning rate as 0.001. $N = 10$.

3.2   Evaluation Criteria

The evaluation is based on the Euclidean distance d between the predicted center of a fovea C and the center of ground truth C', that is $D = d(C, C')$. To obtain a binary output (correct or not), we allow an offset of the predicted foveal center from the center of the ground truth. This default offset is defined as the distance that is $1/2dd$ (1 dd is 80 pixels [17]) and the origin is the foveal center of ground truth. If the predicted foveal centers are within the maximum offset, they are considered as correct, and vice versa. An *accuracy* under the offset contexts is applied for our evaluations (the same as the *success rate* in other literature).

3.3   Results

(1) Evaluation of the Prior Relationship Module

Evaluation on the MF5K data set: As in Table 1, in the MF5K dataset, the accuracy of our fovea localization method is 90.70% when applying prior physiological relationship or 89.93% if not (an accuracy improvement of 0.77%).

Table 1. Comparisons of the proposed fovea localization algorithm using prior physiological relationships (using-prior) or not on the private MF5K dataset. Joint Loc. means the jointly localize the fovea and the optic disc with one Faster R-CNN model.

Methods	Using-prior	Accuracy (%)($D \leq 1/3dd$)
Joint Loc. using faster R-CNN [9]	-	90.23
Faster R-CNN [12]	No	89.93
Our method	Yes	**90.70**

More precisely, for 1301 test images in the MF5K, the accuracy of the optic disc localization is 99.62% (5 images fails). Fovea localization before prior relationship refinement fails in 131 images (an accuracy of 89.93%), among which there are 20 images without any predicted candidate box at all due to severe lesions. Then, after applying the suggested hypothetic centers, 7 (out of 20) images success to infer correct foveal centers within $1/3dd$ offset. Another 3 fundus photographs with unreasonable candidate bounding boxes are removed, and the final outputs are corrected successfully by the relationship prior module.

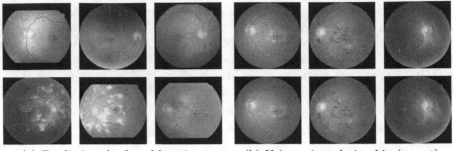

(a) Predicting the foveal location.   (b) Using prior relationship (or not).

Fig. 3. Examples of (a) predicting the fovea bounding boxes. (b) comparison of using prior physiological relationship module (the first row) or not (the second row). (Color figure online)

In addition, for the offset distance D of the predicted results, the mean is $0.16\,dd$, the variance is $0.098\,dd$. The maximum is $4.29\,dd$, and the minimum is 0.

Some examples of the predicted results are shown in Fig. 3(a). The green box is predicted ROI of the fovea and the green cross is the predicted fovea center. The blue cross is the foveal center of the ground truth. Further, three examples using prior relationships or not are compared in Fig. 3(b).

Evaluation on the Messidor data set: While, in the Messidor dataset, due to the relatively better quality of fundus photographs (no serious lesions), each image only has one predicted candidate within the restriction region. Actually, the prior relationship module has not yet been used at all, resulting in no difference between them in terms of accuracy (staying the same as 98.83%).

(2) Evaluation Different Methods on the Public Messidor Dataset
Further, we evaluate our proposed method and the existing methods on the Messidor dataset, as in Table 2. The accuracy of the optic disc localization here is 100%. The accuracy of our method is higher than that of the existing methods with the offset distance as $1/4\,dd$, $1/3\,dd$, and $1/2\,dd$ respectively, which proves the effectiveness of our proposed method. In the case of the $1/2\,dd$ offset, the accuracy of our proposed method is 0.71% higher than the existing method (98.87%) in [2]. In the case of the $1/4\,dd$ offset, the accuracy of our proposed method is 0.67% higher than the existing method (96.83%) in [10].

Table 2. Comparisons of different fovea localization methods in the Messidor dataset.

Methods	Accuracy (%) (D: offset distance)				
	$D \leq \frac{1}{4}dd$	$D \leq \frac{1}{3}dd$	$D \leq \frac{1}{2}dd$	$D \leq dd$	Uncertain
Niemeijer et al. (2009) [10]	96.83	-	97.92	-	
Yu et al. (2011) [16]	95.00	-	-	-	
Gegundez-Arias et al. (2013) [4]	96.08	96.58	96.92	97.83	
Aquino et al. (2014) [1]	91.28	-	98.24	99.56	
Kao et al. (2014) [6]	-	-	97.80		
Dashtbozorg et al. (2016)[2]	93.75	-	98.87	99.58	
Molina-Casado et al. (2017)[8]	96.08	-	98.58	99.50	
Kamble et al. (2017)[5]	-	-	-	-	99.66
Pachade et al. (2019) [11]	-	98.66	-	-	
Our method	**97.50**	**98.83**	**99.58**	**100.0**	

4   Conclusion

In this paper, a simple and effective fovea localization method by the Faster R-CNN with physiological prior in fundus photographs is proposed. The different sizes of the input bounding box of a macular region for training a better Faster R-CNN model are investigated to obtain the optimal input size of the macular bounding box. At the same time, the predicted candidate bounding boxes of the fovea are re-validated with the help of the optic disc location, between which their prior physiological relationships are fully explored and utilized. Experiments on the public Messidor and a private MF5K dataset that is a relatively difficult task show that our proposed method is superior to the state-of-the-art methods, and it improves the prediction accuracy of the fovea localization by 0.71% in the case that the offset distance is less than 1/2 optic disc diameter.

Acknowledgements. This work is supported by the CSC State Scholarship Fund (201806295014), NSFC (No. 61672523, No. 61771468), CAMS Initiative for Innovative Medicine (2018-I2M-AI-001), Beijing Natural Science Foundation (No. 4192029), Beijing Hospitals Authority Youth Programme (QML20170206); The priming scientific research foundation for the junior researcher in Beijing Tongren Hospital, Capital Medical University (2018-YJJ-ZZL-052).

References

1. Aquino, A.: Establishing the macular grading grid by means of fovea centre detection using anatomical-based and visual-based features. Comput. Biol. Med. **55**, 61–73 (2014)
2. Dashtbozorg, B., Zhang, J., Huang, F., ter Haar Romeny, B.M.: Automatic optic disc and fovea detection in retinal images using super-elliptical convergence index filters. In: Campilho, A., Karray, F. (eds.) ICIAR 2016. LNCS, vol. 9730, pp. 697–706. Springer, Cham (2016). https://doi.org/10.1007/978-3-319-41501-7_78

3. Decenciére, E., et al.: Feedback on a publicly distributed image database: the messidor database. Image Anal. Stereology **33**(3), 231–234 (2014)
4. Gegundez-Arias, M.E., et al.: Locating the fovea center position in digital fundus images using thresholding and feature extraction techniques. Comput. Med. Imaging Graph. **37**(5), 386–393 (2013)
5. Kamble, R., et al.: Localization of optic disc and fovea in retinal images using intensity based line scanning analysis. Comput. Biol. Med. **87**, 382–396 (2017)
6. Kao, E.F., et al.: Automated detection of fovea in fundus images based on vessel-free zone and adaptive gaussian template. Comput. Methods Programs Biomed. **117**(2), 92–103 (2014)
7. Lu, S., et al.: Automatic macula detection from retinal images by a line operator. In: ICIP, pp. 4073–4076 (2010)
8. Molina-Casado, J.M., et al.: Fast detection of the main anatomical structures in digital retinal images based on intra- and inter-structure relational knowledge. Comput. Methods Programs Biomed. **149**, 55–68 (2017)
9. Nie, Y., et al.: Joint detection with faster R-CNN. In: ISITC (2016)
10. Niemeijer, M., et al.: Fast detection of the optic disc and fovea in color fundus photographs. Med. Image Anal. **13**(6), 859–870 (2009)
11. Pachade, S., Porwal, P., Kokare, M.: A novel method to detect fovea from color fundus images. In: Iyer, B., Nalbalwar, S.L., Pathak, N.P. (eds.) Computing, Communication and Signal Processing. AISC, vol. 810, pp. 957–965. Springer, Singapore (2019). https://doi.org/10.1007/978-981-13-1513-8_97
12. Ren, S., et al.: Faster R-CNN: towards real-time object detection with region proposal networks. NIPS **1**, 91–99 (2015)
13. Samanta, S., et al.: A simple and fast algorithm to detect the fovea region in fundus retinal image. In: EAIT, pp. 206–209 (2011)
14. Santhi, D., et al.: An efficient approach to locate optic disc center, blood vessels and macula in retinal images. Biomed. Eng. Appl. Basis Commun. **24**(05), 425–434 (2012)
15. Singh, J., et al.: Appearance-based object detection in colour retinal images. In: ICIP, pp. 1432–1435 (2008)
16. Yu, H., et al.: Fast localization of optic disc and fovea in retinal images for eye disease screening. SPIE Med. Imaging Comput.-Aided Diagn. **7963**, 317–328 (2011)
17. Zheng, S.H., et al.: A novel method of macula fovea and optic disk automatic detection for retinal images. J. Electron. Inf. Technol. **36**(11), 2586–2592 (2014)
18. Zheng, Y.: Research on macula area localization of fundus image based on the segmentation of retinal vessel ends. Ph. D. thesis, Huazhong University (2013)
19. Zhou, W., et al.: Detection of macula fovea in a retinal image. J. Image Graph. **23**(3), 442–449 (2018)

Aggressive Posterior Retinopathy of Prematurity Automated Diagnosis via a Deep Convolutional Network

Rugang Zhang[1], Jinfeng Zhao[2], Guozhen Chen[1], Tianfu Wang[1], Guoming Zhang[2(✉)], and Baiying Lei[1(✉)]

[1] National-Regional Key Technology Engineering Laboratory for Medical Ultrasound, Guangdong Key Laboratory for Biomedical Measurements and Ultrasound Imaging, School of Biomedical Engineering, Health Science Center, Shenzhen University, Shenzhen, China
leiby@szu.edu.cn

[2] Shenzhen Eye Hospital, Shenzhen Key Ophthalmic Laboratory, The Second Affiliated Hospital of Jinan University, Shenzhen, China

Abstract. Aggressive Posterior Retinopathy of Prematurity (AP-ROP) is a retinal pathology characterized by severe vasodilation and distortion of the posterior pole of the retina. It may lead to blindness if it is not diagnosed and treated in time. Therefore, early diagnosis of AP-ROP plays a nontrivial role in reducing the blindness rate in children. However, the traditional automated AP-ROP diagnosis methods are based on machine learning with segmentation, where the accuracy is highly dependent on the vessel segmentation. To solve this issue, we propose an approach with two deep convolution networks to automatically diagnose AP-ROP. Specifically, the first network identifies whether the image has the presence of ROP, and the second network divides the ROP images into Regular ROP and AP-ROP. Experimental results show that our proposed method can achieve quite promising AP-ROP diagnosis performance and the transfer learning technique can further boost the automated diagnosis performance.

Keywords: Retinopathy of Prematurity · Automated diagnosis · Deep neural network

1 Introduction

Retinopathy of Prematurity (ROP) is a kind of retinal vascular proliferative blindness, which occurs in premature or low birth weight infant, accounting for about 19% of the causes of blindness in children worldwide. Aggressive Posterior Retinopathy of

This work was supported partly by National Natural Science Foundation of China (Nos. 61871274, 61801305 and 81571758), National Natural Science Foundation of Guangdong Province (No. 2017A030313377), Guangdong Pearl River Talents Plan (2016ZT06S220), Shenzhen Peacock Plan (Nos. KQTD2016053112051497 and KQTD2015033016 104926), and Shenzhen Key Basic Research Project (Nos. JCYJ20170413152804728, JCYJ20180507184647636, JCYJ201708181 42347251, JCYJ20170817112542555 and JCYJ20170818094109846).

H. Fu et al. (Eds.): OMIA 2019, LNCS 11855, pp. 165–172, 2019.
https://doi.org/10.1007/978-3-030-32956-3_20

Prematurity (AP-ROP) is a special type of ROP characterized by plus disease, flat neovascularization (FNV), hemorrhages, and ischemic capillary nonperfusion regions [1, 2]. The condition of AP-ROP is progressing rapidly. If it is not treated in time, the disease will progress to the fifth stage of ROP, and the retina will be completely detached [3]. The progress of AP-ROP is unregular and different from that of Regular ROP [4]. It is easy to cause misdiagnosis and miss the best period of treatment. In addition, the diagnosis of AP-ROP is based on the visual examination by an ophthalmologist, which is subjective. We need a computer-assisted approach to help with early screening of AP-ROP.

Deep learning (DL) has shown an outstanding performance in the field of image analysis [5–7]. It has been successfully used in automatic diagnosis of glaucoma [8, 9], diabetic retinopathy [10], macular degeneration [11] and cataract [12]. It has also been applied to identify ROP with fundus images [13–16]. In addition, it has been used to identify AP-ROP, but the performance is still not satisfactory due to the fact that only a small amount of fundus images is used for the training [3]. More studies are needed to improve the diagnosis accuracy of AP-ROP. This can not only assist the ophthalmologists in early ROP screening of premature infants, but also facilitate the examination procedure and reduce the misdiagnosis rate, which is of great significance in reducing the blindness rate of the disease.

We divide the AP-ROP automatic diagnosis process into two steps. First, we need to determine whether a fundus image is a normal or ROP image, and then divide the ROP image into an AP-ROP or Regular ROP image. Regular ROP is characterized by a slight boundary ridge or demarcation line between the vascularized and no vascularized retina. As shown in Fig. 1(b), a slight white demarcation line appears on the right side of the Regular ROP image, while the characteristics of AP-ROP (Fig. 1(c)) image has occasionally bleeding, vessels dilation and tortuosity around the posterior pole.

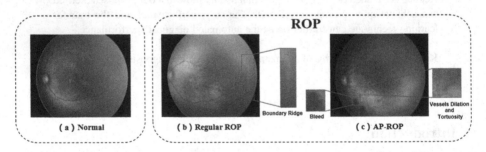

Fig. 1. Fundus image of Normal (a), Regular ROP (b) and AR-ROP (c).

In practice, there are several challenges to be faced:

(1) The ROP dataset includes AP-ROP and Regular ROP, which means that the dataset has several complex features.
(2) The fundus image has a random field of view, so the lesion will appear at different locations in the fundus image.
(3) AP-ROP fundus images are mostly of poor quality.

To accomplish the above challenges, this study devises two advanced classification networks, Network-1 and Network-2 that based on ResNet50 and VGG16. Because used different network can optimized independently based on the task properties to achieve the best diagnosis results.

ResNet50 is a 50-layer deep convolutional neural network [17]. Because it has a deep network layer, it can extract deeper information from the image. That means it can effectively respond to many complex functional problems in the ROP dataset, and is also adaptable to images of poor quality and random field of view. This can solve the complex problem of the data set in the challenge (1). At the same time, we compare the results of various experiments to determine that the network is suitable for Network-1.

VGG16 is a simple and practical network that shows a remarkable performance in image classification tasks [18]. It consists of several consecutive 3 × 3 convolution cores, pooling layers, ReLU layers and fully connected layers. It is one of the most popular convolutional neural network models. The dataset feature complexity in Network-2 is much simpler than that of Network-1. Using the VGG16 network in Network-2 can reduce the loss of information due to the deepening of the network layer while ensuring high accuracy. The final experimental results also show the superiority of VGG16 for Network-2.

The main contributions of this paper are:

(1) We propose a parallel convolutional neural network method to achieve automatic diagnosis of AP-ROP through two tasks.
(2) The convolutional neural networks used by the two tasks are optimized independently based on the task properties to achieve the best diagnosis results.

Our method is evaluated on the self-collected dataset. In the first network, 583 ROP retinal images against a control group of 915 healthy retinal images are used to evaluate the performance of the network. We achieve an accuracy of 96.53%, a sensitivity of 96.74 and a specificity of 96.39 in the first network. In the second network, we evaluate 290 Regular ROP and 293 AP-ROP retinal images, and the results (98.46% accuracy, 100.00% sensitivity and 96.90% specificity) are comparable with the first network. The experimental result demonstrates that the proposed method has achieve remarkable AP-ROP diagnosis performance.

2   Methodology

We train two networks to diagnose AP-ROP automatically. The input of the Network-1 is the ROP disease dataset and normal fundus image dataset, as shown in Fig. 1(a). The Network-1 is used to determine whether the fundus image is a normal or ROP image, while the Network-2 is used to divide the ROP disease data into the Regular ROP fundus images and the AP-ROP fundus images, therefore, we can achieve the purpose of automatic diagnosis of AP-ROP. The flowchart of our proposed method is shown in Fig. 2. In the network training process, we utilize the transfer learning by using ImageNet's pre-training module to obtain prior knowledge from the ImageNet dataset, which not only improves the network's performance, but also speeds up the network learning and convergence. We input the training data set into the network, then extract the features and get the prediction results through the SoftMax classifier.

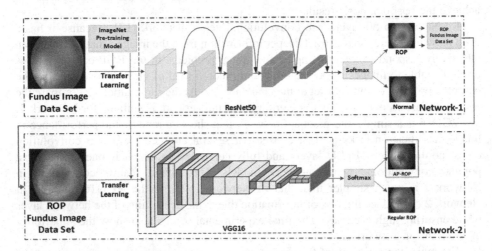

Fig. 2. AP-ROP automatic diagnosis flowchart. The task of the two networks is different. Network-1 diagnoses whether the fundus image is a ROP image. The task of Network-2 is to diagnose whether the ROP fundus image is AP-ROP.

2.1   Architecture of ResNet50

In this paper, Network-1 is based on ResNet50, which is a deep convolutional neural network model and adds a residual module to the network. Residual module allows ResNet50 to solve the gradient disappearance problem of the neural network by learning the difference information between the input and output of the residual module. Because of the existence of the residual module, we can build a deeper neural network, which can extract deeper information of the image and improve the accuracy of the network. When building a neural network, deeper layers mean that deeper feature information can be extracted, but we can't simply overlay the layers of the network. Because the increasing number of network layers will lead to the degradation of the convolutional neural network and even the present gradient vanishing problem. The emergence of ResNet not only solves the problem of degradation of deep neural networks, but also avoids the problem of gradient disappearance of neural networks. In other words, we can use ResNet to build deeper convolutional neural networks.

2.2   Residual Module

In normal network training, if we set the input as x and the output of the network layer is H(x), that is, directly learn as $x \rightarrow H(x)$, it will cause gradient disappear as the number of network layers increases. The ResNet network adds a residual module, which converts the learning function from $x \rightarrow H(x)$ to $x \rightarrow (H(x) - x) + x \rightarrow F(x) + x$ by establishing an identity map. Through the residual module, the neural network does not need to learn the output of the entire network layer, and only needs to learn the residual value of the output between the network layers. This process has not added additional computational resources but can greatly increase the training speed of the network model and improve the final result. Moreover, when the number of network layers is deepened, the residual module can solve the problem of network degradation well. At the same time, the residual module by transforming the 3×3 convolution kernel into two 1×1 convolution kernels, which can effectively reduce the network parameters, greatly decrease the network computing cost, and increase the network learning efficiency.

2.3   Architecture of VGG16

In this study, Network-2 is based on VGG16, which is a bit lighter than Network-1. In this network, the size of the convolution kernel is unified to 3×3. Compared with the large convolution kernel, the small convolution kernel is more advantageous because the multi-nonlinear layer can increase the network depth, which enables that more complex modes can be learned with fewer parameters. But if a network is too deep, it will cause information loss. VGG16 is a moderately deep network, so it can achieve an excellent performance in classification problem.

3   Experiments

3.1   Dataset and Implementation

The RetCam3 data used in this paper is collected by the local hospital. Details of the dataset are listed in Table 1. The data of the Regular ROP includes Stage 1, Stage 2 and Stage 3, and is randomly distributed. In order to save computing resources, we adjust the original fundus image size to $224 \times 224 \times 3$.

Table 1. Dataset distribution in this study.

	Network-1			Network-2		
	ROP	Normal	Total	AP-ROP	Regular ROP	Total
Training set	2676	1300	3976	1419	1257	2676
Test set	583	915	1498	293	290	583
Total	3259	2215	4849	1712	1547	3259

The data of this study is divided into two parts, the training set and the test set. The training set is only used to train the network, and the test set is completely independent to the training set. The test set is used for the performance evaluation of the AP-ROP automatic diagnosis system. The angle of the fundus image data in the test set is random, which can simulate the fundus examination of the newborn in the hospital. The most commonly used evaluation criteria are used to evaluate our model, which consists Accuracy, Sensitivity, Specificity, Precision and F1-score.

3.2   Results

In this study, we compare the VGG16 network with the ResNet50 network, and we also conduct a comparison of transfer learning. The results of Network-1 and Network-2 are shown in Table 2. We can see that ResNet50 has better performance using Network-1, while in Network-2, VGG16 performs better. This is consistent with the conjecture in the introduction. The task of Network-1 is more complicated, thus the deeper network ResNet50 can get better results, while the Network-2 task is much simpler, so the VGG16 is more suitable. The results of the Combined Network are also shown in Table 2. It can be seen that the Network has excellent performance in the diagnosis of AP-ROP.

At the same time, the results of network that using the transfer learning are better than the network that without using it. This shows that transfer learning can effectively optimize the performance of the network and achieve better results.

Table 2. The performance of the networks.

Task	Method	Accuracy	Sensitivity	Specificity	Precision	F1-score
Network-1	**ResNet50+pre-train**	**96.53%**	**96.74%**	**96.39%**	**94.47%**	**95.59%**
	ResNet50	83.44%	79.59%	85.90%	78.25%	78.91%
	VGG16+pre-train	95.46%	94.51%	96.07%	93.87%	94.19%
	VGG16	85.38%	81.30%	87.98%	81.16%	81.23%
Network-2	ResNet50+pre-train	96.40%	96.59%	96.21%	96.26%	96.42%
	ResNet50	95.71%	99.66%	91.72%	92.41%	95.89%
	VGG16+pre-train	**98.46%**	**100.00%**	**96.90%**	**97.02%**	**98.49%**
	VGG16	93.83%	95.90%	91.72%	92.13%	93.98%
Combined network	**ResNet50+VGG16 +pre-train**	**95.93%**	**95.58%**	**97.23%**	**94.29%**	**94.93%**

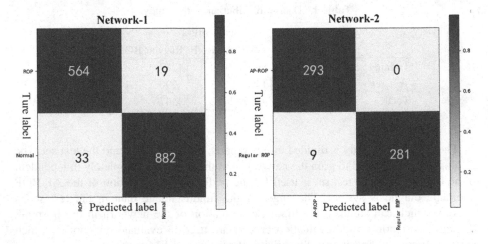

Fig. 3. Confusion matrix for Network-1 and Network-2.

In order to reduce the rate of misdiagnosis, we pay more attention to the false negative rate. For false positives, we re-examine the patient as a diseased sample, but for false negatives, we are likely to miss the sample, this will lead to misdiagnosis. The confusion matrix for Network-1 and Network-2 is shown in Fig. 3. It can be seen that our false negative rate is at a low level, and false negative rate is lower than the false positive rate. This is in line with the hospital's diagnostic needs for AP-ROP, but it will require further testing after actual use in the hospital.

The task of Network-1 is complex. In order to observe the learning status of Network-1, we visualize the feature map of Network-1. Figure 4 shows the feature map of Normal, Regular ROP and AP-ROP. From the feature map of Network-1, we can see that because the Normal image has no clear features, the network has a wide range of attention to the image. In the ROP image, there is a boundary ridge or demarcation line on the Regular ROP image, and the network will predict the image based on this standard. The AP-ROP image is characterized by vessels dilation and tortuosity around the posterior pole. Due to the large range of the posterior pole, it can also be seen that the network has a wide range of attention. These are consistent with medically characterized features of the image.

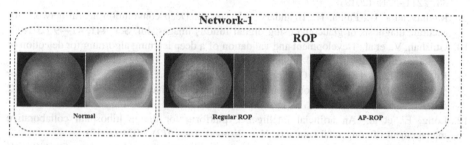

Fig. 4. Feature maps of Network-1 and Network-2. The image characteristics learned by our network are consistent with the information on medical knowledge.

4   Conclusions

In this paper, we use two networks to automatically diagnose AP-ROP in two steps. As can be seen from the test results of this study, our method achieves quite promising performance. In China, many areas lack of experienced ophthalmologists, especially in fundus of premature infants. The incidence of AP-ROP is low and atypical. Many ophthalmologists may have no sufficient experience and are prone to misdiagnosis. If AP-ROP is progressing rapidly, misdiagnosis will miss the best treatment time and lead to serious retinal damage. This study can help doctors to diagnose ROP and classify AP-ROP, which can reduce the misdiagnosis rate of clinicians. The depth convolution network is much faster than humans, and the system can reduce the burden on doctors to a certain extent.

References

1. Vinekar, A., Chidambara, L., Jayadev, C., Sivakumar, M., Webers, C.A., Shetty, B.: Monitoring neovascularization in aggressive posterior retinopathy of prematurity using optical coherence tomography angiography. J. Am. Assoc. Pediatr. Ophthalmol. Strabismus **20**(3), 271–274 (2016)
2. Ahn, Y., et al.: Characteristic clinical features associated with aggressive posterior retinopathy of prematurity. Eye **31**(6), 924–930 (2017)
3. Rajashekar, D., Srinivasa, G., Vinekar, A.: Comprehensive retinal image analysis for aggressive posterior retinopathy of prematurity. PLoS ONE **11**(10), e0163923 (2016)
4. Kim, S.J., et al.: Accuracy and reliability of eye-based vs quadrant-based diagnosis of plus disease in retinopathy of prematurity. JAMA Ophthalmol. **136**(6), 648–655 (2018)
5. Brown, J.M., et al.: Automated diagnosis of plus disease in retinopathy of prematurity using deep convolutional neural networks. JAMA Ophthalmol. **136**(7), 803–810 (2018)
6. LeCun, Y., Bengio, Y., Hinton, G.: Deep learning. Nature **521**(7553), 436–444 (2015)
7. Kermany, D.S., Goldbaum, M., Cai, W., et al.: Identifying medical diagnoses and treatable diseases by image-based deep learning. Cell **172**, 1122–1131 (2018)
8. Diaz-Pinto, A., Colomer, A., Naranjo, V., Morales, S., Xu, Y., Frangi, A.F.: Retinal image synthesis and semi-supervised learning for glaucoma assessment. IEEE Trans. Med. Imaging **38**, 2211–2218 (2019)
9. Raghavendra, U., Fujita, H., Bhandary, S.V., et al.: Deep convolution neural network for accurate diagnosis of glaucoma using digital fundus images. Inf. Sci. **441**, 41–49 (2018)
10. Gulshan, V., et al.: Development and validation of a deep learning algorithm for detection of diabetic retinopathy in retinal fundus photographs. JAMA **316**(22), 2402–2410 (2016)
11. Russakoff, D.B., Lamin, A., Oakley, J.D., Dubis, A.M., Sivaprasad, S.: Deep learning for prediction of AMD progression: a pilot study. Invest. Ophthalmol. Vis. Sci. **60**(2), 712–722 (2019)
12. Long, E., et al.: An artificial intelligence platform for the multihospital collaborative management of congenital cataracts. Nat. Biomed. Eng. **1**(2), 0024 (2017)
13. Coyner, A.S., et al.: Automated fundus image quality assessment in retinopathy of prematurity using deep convolutional neural networks. Ophthalmol. Retina **3**(5), 444–450 (2019)
14. Zhao, J., et al.: A deep learning framework for identifying zone I in RetCam images. IEEE Access **7**, 103530–103537 (2019)
15. Zhang, Y., Wang, L., Wu, Z., et al.: Development of an automated screening system for retinopathy of prematurity using a deep neural network for wide-angle retinal images. IEEE Access **7**, 10232–10241 (2018)
16. Hu, J., Chen, Y., Zhong, J., et al.: Automated analysis for retinopathy of prematurity by deep neural networks. IEEE Trans. Med. Imaging **38**(1), 269–279 (2018)
17. He, K., Zhang, X., Ren, S., Sun, J.: Deep residual learning for image recognition. In: IEEE Conference on Computer Vision and Pattern Recognition, pp. 770–778 (2016)
18. Simonyan, K., Zisserman, A.: Very deep convolutional networks for large-scale image recognition. arXiv preprint, arXiv:1409.1556 (2014)

Automated Stage Analysis of Retinopathy of Prematurity Using Joint Segmentation and Multi-instance Learning

Guozhen Chen[1], Jinfeng Zhao[2], Rugang Zhang[1], Tianfu Wang[1], Guoming Zhang[2($\boxtimes$)], and Baiying Lei[1($\boxtimes$)]

[1] National-Regional Key Technology Engineering Laboratory for Medical Ultrasound, Guangdong Key Laboratory for Biomedical Measurements and Ultrasound Imaging, School of Biomedical Engineering, Health Science Center, Shenzhen University, Shenzhen, China
leiby@szu.edu.cn

[2] Shenzhen Eye Hospital, Shenzhen Key Ophthalmic Laboratory, The Second Affiliated Hospital of Jinan University, Shenzhen, China

Abstract. Retinopathy of prematurity (ROP) is the primary cause of childhood blindness. Prior works have demonstrated the remarkable performances of deep learning (DL) in detecting plus disease and classification between ROP or Normal with retinal images. However, few studies are focused on identifying the "stage" of ROP disease, which is an important factor to evaluate the severity of the disease. In general, only a small region (typical less than 5% of the image) of a fundus image contributes its being classified as different stages of ROP. Therefore, traditional convolutional neural network (CNN) classifier may be ineffective when it is applied to a global feature extraction while the ROP features are localized with a limited number of labeled images. To address this issue, we combine the segmentation and staging, using both fully convolutional network (FCN) and multi-instance learning (MIL) to achieve integrated task of ROP staging and lesions localization. The proposed network is evaluated on 7330 retinal images (2000 Normal, 630 Stage1, 980 Stage2, 870 Stage3 and 250 Stage4) obtained by RetCam3. Experimental results show that the proposed network achieves 0.93 area under the curve (AUC) on the test dataset (accuracy 92.25%, sensitivity 90.53% and specificity 92.35%), and ROP lesions such as demarcation lines, ridges can be accurately located in the fundus images.

Keywords: Multi-instance learning · Segmentation · Retinopathy of prematurity · ROP staging

This work was supported partly by National Natural Science Foundation of China (Nos. 61871274, 61801305 and 81571758), National Natural Science Foundation of Guangdong Province (No. 2017A030313377), Guangdong Pearl River Talents Plan (2016ZT06S220), Shenzhen Peacock Plan (Nos. KQTD2016053112051497 and KQTD2015033016 104926), and Shenzhen Key Basic Research Project (Nos. JCYJ20170413152804728, JCYJ20170817112542555, JCYJ20180507184647636, JCYJ20170818142347251 and JCYJ20170818094109846).

H. Fu et al. (Eds.): OMIA 2019, LNCS 11855, pp. 173–181, 2019.
https://doi.org/10.1007/978-3-030-32956-3_21

1 Introduction

Retinopathy of prematurity (ROP) is a retinal disorder of low birth weight infants and it is the leading cause of childhood blindness [1]. The diagnosis of ROP is based on the retinal fundus images from premature infants. The International Classification of ROP (ICROP), which is developed in 1984 by 23 ophthalmologists from 11 countries, provides a clinical guideline for ROP grading. According to ICROP [2], 5 stages are used to characterize the severity of ROP depending on the appearance of the retinal vessels at the avascular–vascular junction. Stage1 can be identified by the presence of "demarcation line" between the vascularized and no vascularized retina. The demarcation line is relatively flat and has abnormal branching of vessels up to it. Stage2 is characterized by presence of "ridge" in the region of the demarcation line, which is increased in height and width. Stage3 is defined by a ridge with extraretinal fibrovascular proliferation at or just posterior to the ridge. Stage4 has partial retinal detachment and can be further divided into 4A and 4B. Stage5 occurs total retinal detachment. Plus disease, along with the stage of ROP, is defined as venous dilation and arterial tortuosity of posterior pole vessels, which is a type of severe ROP and required early treatment. Plus disease can be found at any stage of ROP. The Stage1, Stage2, Stage3 and Stage4 ROP captured by the RetCam3 are, respectively, shown in Fig. 1.

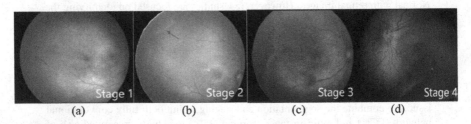

(a)   (b)   (c)   (d)

Fig. 1. Fundus images with different stages of ROP. Stage 1 (a) ROP presents a small demarcation line (red arrow) at the avascular–vascular junction. In stage 2 (b), the width and height of the demarcation line increase and presents a ridge. Stage 3 (c) shows extraretinal fibrovascular proliferation at or just posterior to the ridge and Stage 4 (d) occurs partial retinal detachment. (Color figure online)

In the ROP screening nowadays, fundus images are captured using fundus photography devices, such as RetCam3, and are used by the ophthalmologists to make a diagnosis. A simple screening test and early intervention by an ophthalmologist can prevent the grow of the abnormal vessels, thus prevent the disease turning into blindness [3, 4]. Researches have been done to detect plus disease and aggressive posterior ROP. The multiscale analysis, the semiautomated image analysis software package of retinal images, are used to assess the retinal vessels in ROP based on features such as diameter, tortuosity, and curvature of the vessels. Convolutional neural network (CNN)-based deep learning approach has also been implemented and evaluated for the diagnosis of Normal, pre-Plus, and Plus disease in ROP [5, 6] or to distinguish the images between disease or healthy based on retinal vasculature and achieves quite good performance [7, 8].

However, few studies are focused on identifying the "stage" of ROP, which is important to evaluate the severity of the disease and make appropriate intervention [9]. The reason is that fundus image with label of stage of ROP are scarce and imbalance. In addition, the demarcation lines or ridges presented in the fundus image typically compose only less than 5% of the whole images. Therefore, the traditional CNN classifier may be ineffective when it is applied to a global feature extraction while stage analysis of ROP need to focus on the localized regions of the fundus images [10]. To address the above issues, we propose a joint segmentation and multi-instance learning (MIL) network to improve the classification performance. MIL is firstly introduced with weakly labelled to tackle the problem of drug activity prediction [11] and has been successfully used in detecting Age-Related Macular Degeneration (AMD) with retain images [12]. In MIL setting, the bag is labeled positive if at least one instance of the bag is positive, and the bag is labeled negative if all the instances in it are negative [11]. It is a weakly supervised learning methodology, which means the labels are only assigned to bags of instances. Comparing to the supervised learning, every training instance is assigned with a discrete or real-valued label.

Our proposed framework consists of two main modules, a fully convolutional network (FCN) module and a MIL module. The FCN module serves as end-to-end segmentation and provides a pixelwise binary output, which we call the spatial score map (SSM). The SSM has a size the same as the original image and used as the input of the MIL module. Experimental results verify that the segmentation by the FCN greatly improves the MIL classification performance and the proposed framework achieves the best result in classifying different stages of ROP.

2 Methodology

2.1 Deep Multi-instance Learning Network

The architecture of the proposed model is shown in Fig. 2. The FCN is exploited to extract local feature and perform local disease estimation for a resized fundus image captured from Retcam3. In this study, we use AlexNet as our base network to construct the FCN. An SSM is generated over different lesion regions where each score point of the maps corresponds to disease estimation in the same local regions of original image. Then the estimations of different lesion regions are trained by the MIL network and aggregated to generate the final score.

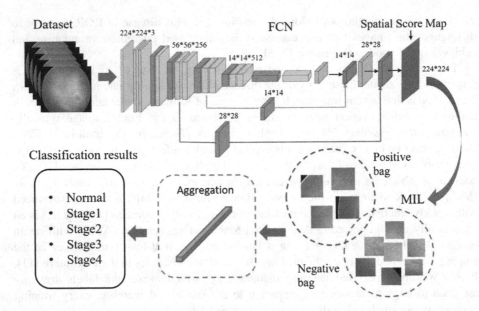

Fig. 2. Architecture of the proposed model. The framework consists of two main modules, a fully convolutional network (FCN) module and a multi-instance learning (MIL) module. The output of the spatial score map (SSM) generated by the FCN is used as the input of the MIL.

2.2   Feature Extraction and Aggregation

CNN has a great representation power of extracting the deep features behind the image. Retaining most blocks of the CNN, this study uses FCN for extracting local features from the fundus image and generating the SSM. In traditional CNN, an input image is downsized and goes through the convolution layers, pooling layer and fully connected (FC) layers, and output one predicted label for the input image. CNN is popular because it can proceed end-to-end training without segmentation, which is a time-consuming task. Instead, if we remove the FC layer of the CNN and perform upsampling, we can gain a pixelwise output, which is the SSM. This is proved to be effective for end-to-end pixel level segmentation [13]. The SSM, instead of the original image, then can be used as the MIL input. To support the MIL assumption, the network is modified a little bit by us. The SSM generated by the FCN is first cropped to 45×45 equal non-overlapping blocks. The SSM is the bag of instance and each block is an instance in it, thus each bag includes 25 instances in this study.

One challenge for the stage of ROP classification lies in the fact that demarcation lines or ridges presented in the fundus image typically compose only a small part of the whole fundus images (as shown in Fig. 3). Therefore, a different aggregation method may give out a tremendous difference of the final result. In this study, supposing $\mathbf{B}_k$ is a bag from the bags set $\{\mathbf{B}_k, k = 1,..., N\}$, then the instance of the bag is $\{\mathbf{B}_{k1},..., \mathbf{B}_{kn}\}$, where n is the number of instances of the bag $\mathbf{B}_k$. The prediction probability of $\mathbf{B}_{kj}$ is denoted as $p_{kj}^c = P(c_{kj} = c|B_{kj})$, where c is the class label and $c \in \{1,\ldots,C\}$. The aggregated function $\mathscr{F}$ for the bag $\mathbf{B}_k$ is:

$$p_k^c = P(c_k = c|B_k) = \mathcal{F}\left(p_{k1}^c,\ldots,p_{kn}^c\right). \tag{1}$$

This study will find out the aggregated function $\mathscr{F}$ that achieves the best performance. Different aggregated function $Avg(p_{kj}^c)$, $Max(p_{kj}^c)$ and Softmax will be performed to compare the model performance. For Softmax, the prediction probability is calculated by:

$$p_k^c = \mathcal{F}\left(p_{k1}^c,\ldots,p_{kn}^c\right) = \frac{\sum_{j=1}^n p_{kj}^c \cdot e^{\alpha p_{kj}^c}}{\sum_{j=1}^n e^{\alpha p_{kj}^c}}, \tag{2}$$

where α is a constance that control the extent to which the Softmax aggregated function approximates a hard max aggregated function.

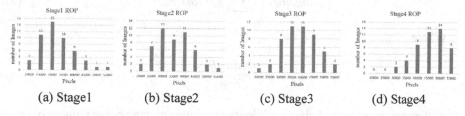

| (a) Stage1 | (b) Stage2 | (c) Stage3 | (d) Stage4 |

Fig. 3. The demarcation lines and ridges presented in the fundus image typically compose less than 5% of the whole fundus images. The size of the original fundus image is 1600×1200, which have a total of 1,920,000 pixels. 50 images of each stage of ROP are selected from the training dataset randomly and the lesion regions are labeled by the ophthalmologists. The total pixels of the lesion regions are calculated automatically by the labeling software.

3 Experiment

3.1 Dataset

Our data includes 1150 ROP examinations from 2016 to 2018 from Eye Hospital. A standard 10-views photograph for an infant's eye is performed in an examination. Thus, every examination consists of 5 to 30 fundus images reflecting to different fundus situation of the whole retina. Images of inadequate quality for clinical diagnosis are removed manually. After that, the data is labeled by three pediatric ophthalmologists in this work, one is senior experts (chief physicians) that have about 20 years of clinical experience in ROP screening and treatment, one is attending physicians that have about 10 years of clinical experience, and the last one is junior ophthalmologist that have about 3 years of clinical experience. We only select images which have consistent label by all the ophthalmologists for the training, and discard ones that have inconsistent label. In total, we get 7330 retinal images from 1073 ROP examinations. We manually separate all fundus images into 5 groups, which are Normal, Satge1, Stage2, Stage3 and Stage4, respectively. Finally, all images are divided into training set, validation set and testing set, respectively. Table 1 shows the data distribution of each dataset in our study.

Table 1. The number of fundus images of different datasets.

	Normal	Stage1	Stage2	Stage3	Stage4
Training set	1600	420	654	580	160
Validation set	200	105	163	145	45
Testing set	200	105	163	145	45
Total	2000	630	980	870	250

From Table 1 we can see that the dataset is imbalanced. The number of Stage4 images are much less than that of Satge1, Stage2 and Stage3. To settle this problem, we introduce median frequency balancing to modify the final loss function, which is formulated as:

$$totalloss = \sum_{1}^{n} \alpha_c \cdot loss(c), \tag{3}$$

Where

$$\alpha_c = \frac{medianfreq}{freq(c)}, \tag{4}$$

α_c denotes coordinate of class c while training, $freq(c)$ denotes the number of class c divided by the total number and $medianfreq$ is the median of these frequencies.

3.2   Classification Results

In this study, the most commonly used evaluation criteria are used to evaluate our classification model, which consists of accuracy (ACC), sensitivity (SEN), specificity (SPEC), area under the curve (AUC) and F1 score (F1, the harmonic mean of precision and recall). For each model, we evaluate the test accuracy for every stage of ROP as well as the total classification accuracy.

Table 2. Accuracy (%) for the proposed model and pure MIL network.

Method	MIL w/o FCN	FCN + MIL		
		Max-agg	Avg-agg	Softmax
Stage1(%)	82.43	88.32	88.76	85.43
Stage2(%)	83.35	93.23	91.07	89.35
Stage3(%)	86.88	91.62	91.14	89.88
Stage4(%)	65.39	74.32	69.54	67.39
Normal (%)	90.71	96.04	93.05	93.71
Total (%)	**86.89**	**92.25**	**90.73**	**90.21**

To save computational resources, the training images are resized from $1600 \times 1200 \times 3$ to $224 \times 224 \times 3$ and the datasets are augmented to four times (rotating each image with a step of 90 degrees). Our method is implemented in the platform of Keras using one NVIDIA TITAN XP GPU with 12 GB RAM. The adaptive moment estimation (Adam) is utilized for optimization and weights update. The learning rate is initially set to 0.0001, then reduced by 0.9 decay when the train loss converges. The weight decay is set to 0.0001 and a mini batch size of 64 is used. To improve model accuracy, all the model training used ImageNet's pre-training module to obtain prior knowledge from the ImageNet dataset, which can also speed up the network learning and convergence. The classification accuracy (%) for every class of the proposed model with different aggregation methods and pure MIL network without FCN are shown in Table 2.

Table 2 shows that the segmentation by the FCN greatly improves the MIL classification performance and the max aggregation obtains a better performance than the average aggregation and Softmax aggregation. We also appraise another five networks, VGG16, AlexNet, Inceptionv4, ResNet50 and ResNet101 to compare the model performance with ours using max aggregation. ResNet50 and ResNet101 have the same network structure, only their depths are different. Experimental results in the test dataset are shown in Table 3.

Table 3. Summary of the model performance in ROP classification in stage.

Method	ACC	SEN	SPEC	AUC	F1
VGG16	86.54	84.25	85.33	0.86	86.30
AlexNet	87.23	88.34	84.57	0.84	83.64
ResNet50	79.47	76.85	80.28	0.81	78.77
ResNet101	75.35	74.28	75.38	0.76	75.40
Inceptionv4	88.17	86.32	90.34	0.90	88.02
Ours	**92.25**	**90.53**	**92.35**	**0.93**	**90.72**

3.3 Visualization Results

The target of this study includes two independent tasks, stage of ROP classification and lesion localization. From the FCN module, the image is input into the AlexNet without the final full connection layer and then fed into the MIL network, and generates the visualized responses from the logistic regression layer. Figure 4 illustrates a case visualization results, among them, the lesion regions of the image predicted to be ROP can be located.

From Fig. 4, we can see that the proposed model is able to learn the lesion region without any explicit bounding box or segmentation ground truth annotation by the ophthalmologist, which greatly relieves the clinical workload. This method can detect the ROP lesions regions in the fundus image, which plays an important role of helping the ophthalmologist for ROP diagnosis and grading.

(a)   (b)   (c)

Fig. 4. The visualization of predicted lesion probabilities in ROP images in Stage1 (a), Stage2 (b) and Stage3 (c).

3.4 Discussion

Form Table 3, we can see that the proposed method achieves the best accuracy. The FCN + MIL network improves greatly over the baseline models (e.g., AlexNet, VGG, ResNet50, ResNet101 and Inceptionv4), which implies that the SSM generated by the FCN successfully extracts the high-level information of the fundus images. Using the SSM rather than the original image for the MIL training can provide better classification results. However, there are still some wrong predictions, specially appearing in Stage4 ROP prediction. After the test, a total of 9 Stage4 images are misclassified as Normal due to the vagueness in the characteristics of Stage4 images, which implies that the FCN may not working well in extracting a global information.

This study also has several limitations. First, Plus disease is excluded from this study in order to focus on identification of the different stages of ROP. Stage5 ROP is also excluded since it is rare and clear retain image is not available for training as the retina is totally detached. Second, the accuracy of any artificial intelligence system is dependent on the quality of data it presents. In this study, images of inadequate quality for clinical diagnosis are excluded manually. In the future, we will develop a software to automatically determine if images are of sufficient quality.

4 Conclusion

We propose a joint segmentation and multi-instance learning network to effectively classify the ROP images in stage as well as to localize the ROP lesions in fundus images. Two independent tasks are achieved in one model. Results on the test dataset show that our model outperforms conventional CNN-based classifier under the same number of parameters in deep model. The method can detect the ROP lesions such as demarcation lines and ridges, which plays an important role in ROP diagnosis and grading.

References

1. Hellström, A., Smith, L.E.H., Dammann, O.: Retinopathy of prematurity. The Lancet **382** (9902), 1445–1457 (2013)
2. International Committee for the Classification of Retinopathy of Prematurity (2005).: The international classification of retinopathy of prematurity revisited. Archives of Ophthalmology, 123(7), 991–999 (2005)

3. Rao, J., et al.: Trend and risk factors of low birth weight and macrosomia in south China, 2005–2017: a retrospective observational study. Sci. Rep. **8**(1), 1–8 (2018)
4. Quinn, G.E., et al.: Incidence and early course of retinopathy of prematurity: secondary analysis of the postnatal growth and retinopathy of prematurity (G-ROP) study. JAMA Ophthalmol. **136**, 1383–1389 (2018)
5. Brown, J.M., et al.: Automated diagnosis of plus disease in retinopathy of prematurity using deep convolutional neural networks. JAMA Ophthalmol. **136**, 803–810 (2018)
6. Hu, J., Chen, Y., Zhong, J., Ju, R., Yi, Z.: Automated analysis for retinopathy of prematurity by deep neural networks. IEEE Trans. Med. Imaging **38**, 269–279 (2018)
7. Zhang, Y., et al.: Development of an automated screening system for retinopathy of prematurity using a deep neural network for wide-angle retinal images. IEEE Access **7**, 10232–10241 (2018)
8. Ting, D.S., Wu, W.-C., Toth, C.: Deep learning for retinopathy of prematurity screening. Br. J. Ophthalmol. **103**, 580–584 (2018)
9. Kim, S., Port, A.D., Swan, R., Campbell, J.P., Chan, R., Chiang, M.: Retinopathy of prematurity: a review of risk factors and their clinical significance. Surv. Ophthalmol. **63**, 618–637 (2018)
10. Lu, J., Hu, J., Zhao, G., Mei, F., Zhang, C.J.C.: An in-field automatic wheat disease diagnosis system. Comput. Electron. Agric. **142**, 369–379 (2017)
11. Dietterich, T.G., Lathrop, R.H., Lozano-Pérez, T.: Solving the multiple instance problem with axis-parallel rectangles. Artif. Intell. **89**, 31–71 (1997)
12. Liu, H., Wong, D.W., Fu, H., Xu, Y., Liu, J.: DeepAMD: detect early age-related macular degeneration by applying deep learning in a multiple instance learning framework. ACCV **5**, 625–640 (2019)
13. Long, J., Shelhamer, E., Darrell, T.: Fully convolutional networks for semantic segmentation. In: Proceedings of the IEEE conference on computer vision and pattern recognition, pp. 3431–3440 (2015)

Retinopathy Diagnosis Using Semi-supervised Multi-channel Generative Adversarial Network

Yingpeng Xie[1], Qiwei Wan[1], Guozhen Chen[1], Yanwu Xu[2], and Baiying Lei[1(✉)]

[1] National-Regional Key Technology Engineering Laboratory for Medical Ultrasound, Guangdong Key Laboratory for Biomedical Measurements and Ultrasound Imaging, School of Biomedical Engineering, Health Science Center, Shenzhen University, Shenzhen, China
leiby@szu.edu.cn

[2] Ningbo Institute of Industrial Technology, Chinese Academy of Sciences, Ningbo, China

Abstract. Various kinds of retinopathy are the leading causes of blindness in human being, and with the rapid development of fundus images (FI) analysis in recent years, deep learning has became the focus while using Computer-Aided-Diagnosis (CAD) system to diagnose retinopathy. However, the conventional deep learning method usually rely on sufficient number of labeled FI, but the cost of acquiring enough labeled FI is too expensive, in addition the difficulty in identifying the tiny lesions and the randomness of lesions distribution also brings great challenge to CAD via deep learning. Therefore, this paper proposes a semi-supervised multi-channel generative adversarial network (GAN), which can reasonably utilize the unlabeled images and generate new samples to reduce the dependence on the labeled images, meanwhile we introduce into a general feature extraction strategy to avoid the problem of image valid information disappearance caused by downsampling, and improve the robustness of the generative and discriminant network. The experimental results show that our proposed network boosts the classification accuracy by 10.5% compared with the control network using conventional method and reaches the highest of 88.9%.

Keywords: Retinopathy · Semi-supervised · Multi-channel · Generative adversarial network · General feature extraction

This work was supported partly by National Natural Science Foundation of China (Nos. 61871274, 61801305 and 81571758), National Natural Science Foundation of Guangdong Province (No. 2017A030313377), Guangdong Pearl River Talents Plan (2016ZT06S220), Shenzhen Peacock Plan (Nos. KQTD2016053112051497 and KQTD2015033016 104926), and Shenzhen Key Basic Research Project (Nos. JCYJ2017 0413152804728, JCYJ20180507184647636, JCYJ20170818142347251 and JCYJ20170818094109846).

H. Fu et al. (Eds.): OMIA 2019, LNCS 11855, pp. 182–190, 2019.
https://doi.org/10.1007/978-3-030-32956-3_22

1 Introduction

The eyes are the organs of human that distribute many abundant and dense microvascular, various kinds of unhealthy states of the human body may cause direct or indirect harm to the eyes. According to the WHO statistics on blindness and visual impairment [1], at present about 1.3 billion people worldwide suffer from different forms of vision impairment, of which about 217 million people suffer from moderate to severe far-vision impairment due to pathological causes, which mainly include pathological myopia (PM) caused by ametropia, glaucoma induced by high intraocular pressure, cataract caused by protein degeneration of lens, age-related macular degeneration (AMD) caused by vascular sclerosis, and diabetic retinopathy (DR) induced by blood glucose fluctuation.

At the present stage, the diagnosis of retinopathy still rely on the professional ophthalmologists, which are not only costly and laborious, but also prone to delay the treatment of patients, so it is in urgent need of an automatic auxiliary diagnosis method. At present, in order to improve the efficiency of regular examination, advanced fundus imaging technology (FIT) has been widely used in most medical institutions, these high resolution fundus images (HRFI) provide the possibility of CAD applied in retinopathy diagnosis. However, in traditional machine learning, most of effective lesions features information needs to be extracted manually by specialist, and the final performance often depends on the reliability of such extracted features [12], it greatly depends on the clinical prior knowledge. Fortunately at present deep learning method can skip this manual feature extraction step, automatically extract effective features of image, which presents more stable and excellent performance.

Although the existing FIT can obtain fundus images at a relatively low cost and ensure the resolution and clarity, but due to personal privacy and labeling reliability, labeled FI are still extremely scarce compared with natural data sets and a surplus of unlabeled FI are also usually produced, so it is unsuitable to use usual deep learning method [9]. According to above analysis, we think it is necessary to introduce semi-supervised learning, and as we know, the GAN not only can be used to generate image [4,13], but also show an excellent ability in semi-supervised classification [10]. In addition, FI obtained by current FIT always has a fairly high resolution, which is necessary to avoid losing the information of tiny lesions for clinical diagnosis, but we have to face some challenges when using such HRFI as network input: (1) The discriminator needs a corresponding depth to downsample the effective information. (2) The generator should be able to generate such high resolution fundus images containing valid semantic information. Therefore we put the pre-trained deep convolutional network, which can extract the general features [11] of image, at the front end of our network to process the HRFI, thus compressing the network input with low distortion.

To sum up, we propose a feature extraction based multi-channel GAN with semi-supervision, the main contributions can be summarized as follows:

- The feature extractor can compress the input images with low distortion, alleviating the dispersion problem of tiny lesions information.

- Semi-supervised learning is introduced in GAN, which can utilize the unlabeled images to help our network learn extra effective information.
- Multi-channel generator is configured to cooperatively generate new samples, which addresses the issue of the deficiency of training images.

2   General Feature Extraction Based Multi-channel GAN with Semi-supervision

2.1   Theoretical Basis

Feature Extraction. In deep learning, we usually adopt transfer learning [11] in the case of insufficient images to avoid overfit, pre-trained deep convolutional networks are usually adopted to extract general features of images, and only simple softmax/support vector machine classifiers are trained to classify. Based on this idea, it is easy to think that we can employ such a network to extract HRFI general feature, thus compressing our network input, which not only greatly reduces the parameters of the network, but also makes the generator easier to converge effectively, the intuition is that generator generators generate feature vectors instead of generating high-resolution images.

Semi-supervised Learning. Considering a simple supervised K classification question, set the input image as x_i, and the output of classifier is an unnormalized logarithmic probability vector $(o_1, o_2,..., o_K)$, if softmax classifier is adopted, the normalized probability vector can be obtained by Softmax function, then the classifier's parameter can be updated to minimize the cross entropy loss between it and the corresponding label though back propagation.

$$Li = -\log\left(\frac{e^{o_{y_i}}}{\sum_K e^{o_k}}\right) \tag{1}$$

When we need to consider the images with much of unlabeled, combining the idea of GAN, the strategy is to exploit the classifier inherit general discriminator's ability, namely the classifier is also able to identify whether the input image is the fake image generated by the generator or the real image in our training dataset, so that it becomes a K + 1 classification problem, thus the unlabeled images have a temporary label, that is, the real image [14].

Generative Adversarial Network. GAN [6] was originally a machine learning model, which can learn the potential distribution of high dimensional data and generate similar data. Nowadays, GAN is usually composed of a deconvolution network called generator and a convolution network called discriminator, the purpose of the generator is to generate fake data $G(z)$ by using noise data z to fit the real data x statistical distribution, while discriminator is to distinguish the real and fake data. In game theory, this competition is called the minimax game, in which the generator repeatedly tries to fool the discriminator, which tries to see through the generator's tricks to distinguish the real data from fake.

$$\min_G \max_D \mathbb{E}_{x \sim data}[\log D(x)] + \mathbb{E}_{z \sim noise}[\log(1 - D(G(z)))] \tag{2}$$

where $D(x) = 1 - P_{model}(y = K + 1|x)$, and $P_{model}(y = K + 1|x)$ represents the probability that the discriminator determines that x is a fake data, so $D(x)$ is the probability that the discriminator says x is real.

2.2 Materials and Methods

Dataset. We collect images from multiple databases while ensuring quality, which ensures the network have generalization effect on FI obtained by different shooting equipments, technologies and environment in practical application. The detailed description of experimental data is shown in Table 1, then we divided labeled images into 70% training data, 10% validation data and 20% test data, while all unlabeled images are used as training data.

Table 1. Datasets

Classification	DataSource	Amount
Normal	ISBI 2019 Palm(184) + HRF(15)[2] + iChallenge-AMD(195) + Diaretdbv0(20)[8]	414
PM	ISBI 2019 Palm	213
Glaucoma	MICCAI 2018 REFUGE(40) + HRF(15) + Share(97)	152
Cataract	Share	75
AMD	iChallenge-AMD	87
DR	Diaretdbv0(110) + HEI-MED(169)[5] + HRF(15)	294
Unlabeled	Diaretdbv1(89)[7] + ISBI 2019 Palm(400) + Share(276) + MICCAI 2018 REFUGE(1153)	1918

Image Preprocessing. The black boundary region is a common classification disturbance and need to be removed first, then three image augmentation techniques [3], namely Random-Brightness-Contrast, Random-Gamma and CLAHE, are used in combination to improve image quality intuitively by eliminating blurring and adjusting brightness imbalance. It is worth mentioning that although images in each database have different resolutions, the feature vetor of the uniform size will be obtained by feature extractor which supports for variable input size, therefore no extra interpolation operation is applied to the images.

Semi-supervised Multi-channel Generative Adversarial Network

Configuration. To put it simply, we only adjusts the following strategies on the general GAN to construct our network, the framework is shown in Fig. 1.

- Feature extractor are employed to process the image to extract feature vetor, resulting in uniform size network input.

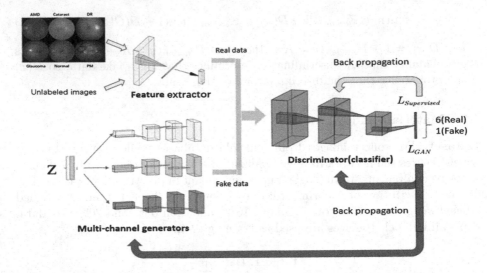

Fig. 1. Multi-channel semi-supervised generative adversarial network

- Training data includes labeled images, unlabeled images and generated images. The former two are jointly labeled as the real image and the last as the fake image alone.
- Multi-channel generator with different channel configurations work together to generate new samples, and the discriminator should correctly classify labeled images on the basis of distinguishing the real image from fake.

Loss Function. Based on the analysis on the theoretical basis, the loss function of our network is divided into two parts: the cross entropy loss of the supervised network and the unsupervised game loss of the GAN [15].

$$L = L_{\text{Supervised}} + L_{\text{GAN}} \tag{3}$$

where:

$$L_{\text{Supervised}} = -\mathbb{E}_{x \sim p_{\text{data}}} \log p_{\text{model}}(y|x, y < K+1)$$
$$L_{\text{GAN}} = L_{\text{Classifier}} + L_{\text{Generator}}$$
$$= -\{\mathbb{E}_{x \sim p_{data}} \log[D(x)]$$
$$+ \mathbb{E}_{z \sim noise} \log[(1 - D(G(z)))]\}$$
$$- \mathbb{E}_{z \sim noise} \log[D(G(z))]$$

The target of loss function is as follows: (1) Supervised network: Discriminator should have the ability to classify labeled images as corresponding label category. (2) GAN: Discriminator should identify labeled and unlabeled images both as the real image and identify generated images as the fake image, meanwhile generator should generate image which is closer to the real image to interfere with the discriminator.

3   Experiments and Results

Details. We employ Inception_Resnet_V2 [16] trained on ImageNet (ILSVRC-2012-CLS) as feature extractor, the output for each image is a feature vector of length 1536, which then is resized to 32×48 and normalized to $[-1, 1]$, then one-sided label smoothing ($\alpha = 0.9$) is applied to the positive label [17]. For the network, we configure eight-channel generator and seven-classification discriminator, both of which adopt Adam optimizer to train 30000 iterations with same first-order momentum of 0.5 and the learning rate of 0.0001 and 0.00008 respectively, besides the generator updates twice as frequently as the discriminator.

Experimental Settings. In order to better analyze our method, we use the existing labeled images to fine tune Inception_Resnet_V2 to set a control network, then our model is evaluated qualitatively and quantitatively by comparing with the control network with different hyper-parameter, that is, whether to use image augmentation (RAW/AUG) and whether to include unlabeled images for training (Unlabeled/No unlabeled).

Qualitative Analysis. Figure 2 shows the qualitative result. It can be seen that compared with the control network, the fitting accuracy result (Deep color curve) of our network under the same training times has significant accuracy improvement and shows an upward trend, meanwhile the fluctuation range of original result (Light color curve) is decreasing, which shows that our network indeed has more excellent and stable classification performance.

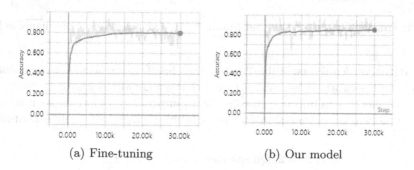

(a) Fine-tuning   (b) Our model

Fig. 2. Accuracy on the validation set in the training process (Color figure online)

Data dimension reduction is a method commonly used for visual high-dimensional data, which can intuitively reveal the potential connections of high-dimensional data. In this study, t-SNE is adopted, Fig. 3 presents the visualization of general feature vetors extracted from the training set and the test set by feature extractor and the corresponding outputs of our network. It is observed from Fig. 3(a,d) that general feature vetors has been able to reflect the potential connections among some same category of data, that is, although all the points are in one block, there are some color blocks. Then from Fig. 3(b,e) and (c,f) we can see the overfitting phenomenon, this is not for nothing, which is determined by the nature of generative network, because the information that the generator can access and learn is only from the training set, and in fact, the generator's role is precisely to enhance the discriminator's learning of these information, but there is no need to worry too much, as unlabeled data increases, more extra information will benefit to generator, reducing the influence of overfitting.

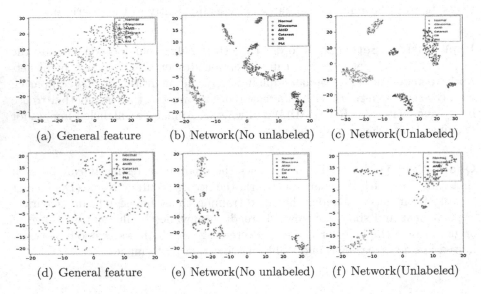

(a) General feature   (b) Network(No unlabeled)   (c) Network(Unlabeled)

(d) General feature   (e) Network(No unlabeled)   (f) Network(Unlabeled)

Fig. 3. Training set (up) and test set (down) feature visualization (Color figure online)

Quantitative Analysis. Table 2 shows the classification accuracy of our model and control network, it is obvious to see that image augmentation technique is a general method to improve classification performance, what really matters is that our networks can indeed utilize the unlabeled images to benefit discriminant

Table 2. Quantitative analysis on test set

Method	Accuracy/%
Fine-tuning (RAW + No unlabeled)	78.4
Fine-tuning (AUG + No unlabeled)	80.2
Our method (RAW + No unlabeled)	86.7
Our method (AUG + No unlabeled)	87.5
Our method (RAW + Unlabeled)	87.8
Our method (AUG + Unlabeled)	88.9

network, although the effect shown on the table may seem not so significant, but there are often far more unlabeled images than the labeled in practice, thus we can expect more improvement in the further work.

4   Conclusion

Faced with the common problem of the scarcity of labeled images in the medical field, our proposed semi-supervised multi-channel GAN can generate new samples and reasonably utilize unlabeled images to help complete retinopathy diagnosis. The main contributions of this project are as follows: (1) A multi-channel GAN is proposed, which can generate fake images containing representative information of existing images to enhance the discriminator's learning. (2) The data of multiple databases are used to simulate the actual scene. (3) Feature extractor can extract effective features of HRFI, resulting in small size network input, which significantly reduces the parameters of discriminator and makes the generator easier to work. Compared with the conventional supervised deep learning method, our network has more excellent and stable classification performance, finally reaches the diagnostic accuracy of 88.9%. In practice, our network's advantages can be taken to have considerable development potential in the deep learning problems when there are difficult to collect data and have high labeling cost.

References

1. Bourne, R.R., et al.: Magnitude, temporal trends, and projections of the global prevalence of blindness and distance and near vision impairment: a systematic review and meta-analysis. Lancet Glob. Health **5**(9), e888–e897 (2017)
2. Budai, A., Bock, R., Maier, A., Hornegger, J., Michelson, G.: Robust vessel segmentation in fundus images. Int. J. Biomed. Imaging **2013** (2013)
3. Buslaev, A., Parinov, A., Khvedchenya, E., Iglovikov, V.I., Kalinin, A.A.: Albumentations: fast and flexible image augmentations. ArXiv e-prints (2018)
4. Denton, E.L., Chintala, S., Fergus, R., et al.: Deep generative image models using a laplacian pyramid of adversarial networks. In: Advances in Neural Information Processing Systems, pp. 1486–1494 (2015)
5. Giancardo, L., et al.: Exudate-based diabetic macular edema detection in fundus images using publicly available datasets. Med. Image Anal. **16**(1), 216–226 (2012)
6. Goodfellow, I.J., et al.: Generative adversarial nets. In: International Conference on Neural Information Processing Systems (2014)
7. Kauppi, T., et al.: The DIARETDB1 diabetic retinopathy database and evaluation protocol. BMVC **1**, 1–10 (2007)
8. Kauppi, T., et al.: DIARETDB0: evaluation database and methodology for diabetic retinopathy algorithms. Mach. Vis. Pattern Recogn. Res. Group Lappeenranta Univ. Technol. Finland **73**, 1–17 (2006)
9. Litjens, G., et al.: A survey on deep learning in medical image analysis. Med. Image Anal. **42**, 60–88 (2017)
10. Odena, A.: Semi-supervised learning with generative adversarial networks. arXiv preprint. arXiv:1606.01583 (2016)

11. Pan, S.J., Yang, Q.: A survey on transfer learning. IEEE Trans. Knowl. Data Eng. **22**(10), 1345–1359 (2009)
12. Priya, R., Aruna, P.: Diagnosis of diabetic retinopathy using machine learning techniques. ICTACT J. Soft Comput. **3**(4), 563–575 (2013)
13. Radford, A., Metz, L., Chintala, S.: Unsupervised representation learning with deep convolutional generative adversarial networks. arXiv preprint. arXiv:1511.06434 (2015)
14. Salimans, T., Goodfellow, I., Zaremba, W., Cheung, V., Radford, A., Chen, X.: Improved techniques for training GANs. In: Advances in Neural Information Processing Systems, pp. 2234–2242 (2016)
15. Sutskever, I., Jozefowicz, R., Gregor, K., Rezende, D., Lillicrap, T., Vinyals, O.: Towards principled unsupervised learning. arXiv preprint. arXiv:1511.06440 (2015)
16. Szegedy, C., Ioffe, S., Vanhoucke, V., Alemi, A.A.: Inception-v4, inception-ResNet and the impact of residual connections on learning. In: Thirty-First AAAI Conference on Artificial Intelligence (2017)
17. Warde-Farley, D., Goodfellow, I.: 11 adversarial perturbations of deep neural networks. Perturbations Optim. Stat. **311** (2016)

Author Index

Printed in the United States
By Bookmasters